PROBABILITY: A FIRST COURSE

PROBABILITY: A FIRST COURSE

IAN S. MURPHY

Department of Mathematics
University of Glasgow

ARKLAY PUBLISHERS

© Ian S. Murphy, 1991.

ISBN 0 9507126 6 3

First Edition 1991.

All rights reserved. No part of this work may be reproduced, stored in any information storage and retrieval system, or transmitted in any form or by any means, electronic or mechanical, including photocopying or recording, without the prior permission in writing of the copyright owner.

Published by Arklay Publishers,
64 Murray Place, Stirling, Scotland. FK8 2BX.

Printed and Bound in Great Britain by Bell and Bain Ltd., Glasgow

PREFACE

This book is primarily intended for students attending a first course in Probability. Many people attending such courses say that Probability is hard: this book aims to provide such people with a lifeboat.

It seems to me that the key to success in Probability is the ability to think clearly: there are various standard types of problem and it is generally better to think how you can relate the problem in hand to one of the standard types rather than to lash out with the first ad hoc argument that springs to mind. At the start it is vitally important that you understand the ideas of sample space (§3) and event (§4 and §5): without these you will always be vulnerable. If you are thinking clearly you will realise that you also have to choose your words carefully when you write things down. In this connection, particular attention should be paid to the words <u>or</u> and <u>and</u> (§1). For my part, I have tried to produce a cleanness about the words in the book, particularly in the examples.

Though the book is primarily aimed at students attending lectures, I know that it is possible to have opinions about and interest in Probability without any formal mathematical background at all. If you are reading the book from this point of view I would make the following three points to you. First, don't imagine you have to read the book in the order of the chapters given: you can read pieces wherever you like. Secondly, you will have to address the ideas of binomial coefficient notation (§11) and set theory (§1) fairly early on, but these are not major obstacles. Thirdly, Calculus in the form of integration, appears in Chapters 7, 8, 10 and 11, but you can do a good part of Chapter 8 and almost all of Chapter 10 without knowing how to integrate.

The book is based on experience with a basic Probability course given to students of Science and Engineering over a number of years here in the University of Glasgow.

Finally, I wish to record my thanks to my colleague, Dr. Neil K. Dickson, who very kindly checked the completed text in detail: on various points his thinking has been clearer than my own.

Ian S. Murphy

University of Glasgow

May 1991

CONTENTS

Chapter		Page
1.	Basic Ideas	1
2.	Counting cases, Equally likely outcomes, The Hypergeometric Distribution	9
3.	Conditional Probability	18
4.	Independence	27
5.	One Dimensional Random Variables in General	35
6.	The Binomial Distribution	37
7.	Continuous Random Variables	44
8.	Expected Value and Variance of a Random Variable	52
9.	The Poisson Distribution	68
10.	The Normal Distribution	79
11.	Two Dimensional Random Variables	90
12.	Chebyshev's Inequality and the Central Limit Theorem	98
	Appendix on the Gamma Function (§100)	108
	Hints and Answers to the Examples	110
	Table of the Φ function for the Normal Distribution	118
	Index	119

I. BASIC IDEAS

1. A RECAP ON SET THEORY

EXAMPLE. Let $S = \{1, 2, 3, 4, 5, 6, 7, 8, 9\}$.
Let A and B be the subsets of S given by

$$A = \{1, 3, 5, 7, 9\} \text{ and } B = \{1, 2, 3, 4\}.$$

Find the sets $A \cup B$, $A \cap B$, $\overline{A} \cup \overline{B}$ and $\overline{A} \cap \overline{B}$, where $\overline{A}$ denotes the complement of A.

Solution. $A \cup B = \{1, 2, 3, 4, 5, 7, 9\}$.
$A \cap B = \{1, 3\}$.
Also $\overline{A} = \{2, 4, 6, 8\}$ and $\overline{B} = \{5, 6, 7, 8, 9\}$.
So $\overline{A} \cup \overline{B} = \{2, 4, 5, 6, 7, 8, 9\}$ and $\overline{A} \cap \overline{B} = \{6, 8\}$.

Notice: (i) The words OR, AND and NOT are associated with the operations of union, intersection and complement as illustrated in the following diagrams.

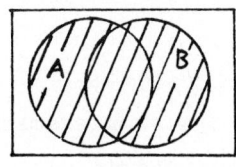

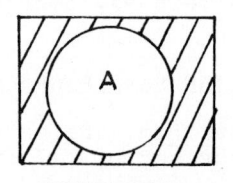

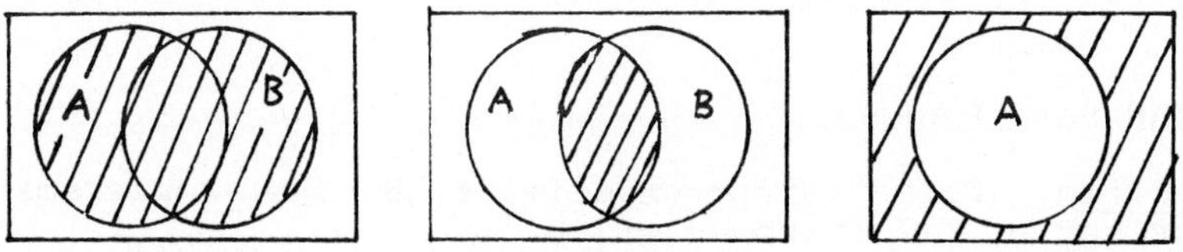

UNION	INTERSECTION	COMPLEMENT
$x \in A \cup B$	$x \in A \cap B$	$x \in \overline{A}$
$x \in A$ or $x \in B$	$x \in A$ and $x \in B$	x not in A

[Note that "$x \in A$" means "x is a member of set A".]

(ii) The following standard rules apply to all sets A, B, C:

$A \cup (B \cap C) = (A \cup B) \cap (A \cup C);$
$A \cap (B \cup C) = (A \cap B) \cup (A \cap C);$ (Distributive Laws)

$\overline{A \cup B} = \overline{A} \cap \overline{B};$
$\overline{A \cap B} = \overline{A} \cup \overline{B}.$ (De Morgan's Laws)

We make occasional reference to these later. They can be checked using Venn diagrams (e.g. Ex. 1.10) but do not spend long looking at them here.

(iii) The symbol $\emptyset$ is used to denote the empty set, i.e. the set with no member.

EXAMPLES TO DO: Page 7: Ex. 1.

2. RANDOM EXPERIMENTS

A <u>random experiment</u> is an experiment the outcome of which cannot be predicted beforehand. We give below three examples of random experiments here. We will refer to these in §3 and subsequently.

<u>EXPERIMENT 1</u>. A coin is tossed three times and the <u>total number</u> of heads is noted.

<u>EXPERIMENT 2</u>. A coin is tossed three times and the <u>sequence of</u> heads and tails is noted.

<u>EXPERIMENT 3</u>. A light bulb is bought and is burnt until <u>it burns out</u>. Its total life length in hours is noted.

<u>Notice</u>: (i) In each experiment the outcome cannot be predicted beforehand.

(ii) Not only the experimental details are given but also what is being observed.

(iii) If each experiment is repeated many times a pattern may emerge.

3. THE SAMPLE SPACE

<u>Definition</u>. For a random experiment E the <u>sample space</u> is the set of all outcomes of E.

<u>EXAMPLE</u>. Describe a sample space for each of the experiments in §2.

<u>Solution</u>. For experiment 1 we can take the sample space

$$S_1 = \{0, 1, 2, 3\}.$$

For experiment 2 we can take the sample space

$$S_2 = \{HHH, HHT, HTH, THH, HTT, THT, TTH, TTT\}.$$

For experiment 3 we can take the sample space

$$S_3 = \{x : x \text{ a real number}, x \geq 0\}.$$

<u>Notice</u>: (i) The sample space is the set of all possible outcomes, i.e. the set of all possible results that an observer could write in his notebook. As such it can also be called the <u>record set</u> or <u>outcome set</u>, names which have the merit of <u>emphasising</u> that <u>it is a set</u>.

(ii) The experimental details of experiments 1 and 2 are similar but notice that <u>different things are being observed</u>. It is this that gives different sample spaces.

For any random experiment it is vital to be absolutely clear about <u>what is being observed</u>.

    (iii) Actually, S_2 <u>is</u> acceptable as a sample space for experiment 1. An observer in experiment 1 who decides to record his results as HHH, HHT, etc is preserving enough information to be able to say how many heads have occurred (as he is required to do by the experiment). On the other hand, S_1 is <u>not</u> acceptable as a sample space for experiment 2, because an observer in experiment 2 who records just the total number of heads has lost information which is required to say which sequence of heads and tails has occurred (as he is required to do by the experiment).

<u>EXAMPLES TO DO</u>: Page 7: Exs. 2, 3, 4.

4. EVENTS

<u>Definition</u>. Let E be a random experiment with sample space S. Then an <u>event</u> A is any subset of S.

<u>Notice</u>: Any subset of S is an event. Usually we describe events in words, but keep in mind that the description corresponds to some subset of S. The following example illustrates this.

<u>EXAMPLE</u>. Every day two runners pass a house - one a man, the other a woman. An observer in the house records the times at which they pass (on the scale $0 < x \leq 24$ in the twenty four hour clock system), the times being x for the man and y for the woman. Describe a sample space for this experiment and give a representation of it in the xy-plane. Sketch the following events in the xy-plane showing each on a separate diagram:

    (a) the man passes before the woman,

    (b) the two runners pass at the same time,

    (c) the woman passes at least 6 hours before the man,

    (d) there is less than three hours' gap between the two runners,

    (e) the man passes before 0900 <u>and</u> the woman passes after 1600,

    (f) the man passes before 0900 <u>or</u> the woman passes after 1600.

<u>Solution</u>. We can take the sample space as $S = \{(x,y) : 0 < x \leq 24, 0 < y \leq 24\}$. In the xy-plane we can represent this as a square as shown. (The edges of the square on the x- and y-axes are not included here because x and y cannot be 0.)

The various events can then be drawn as follows:

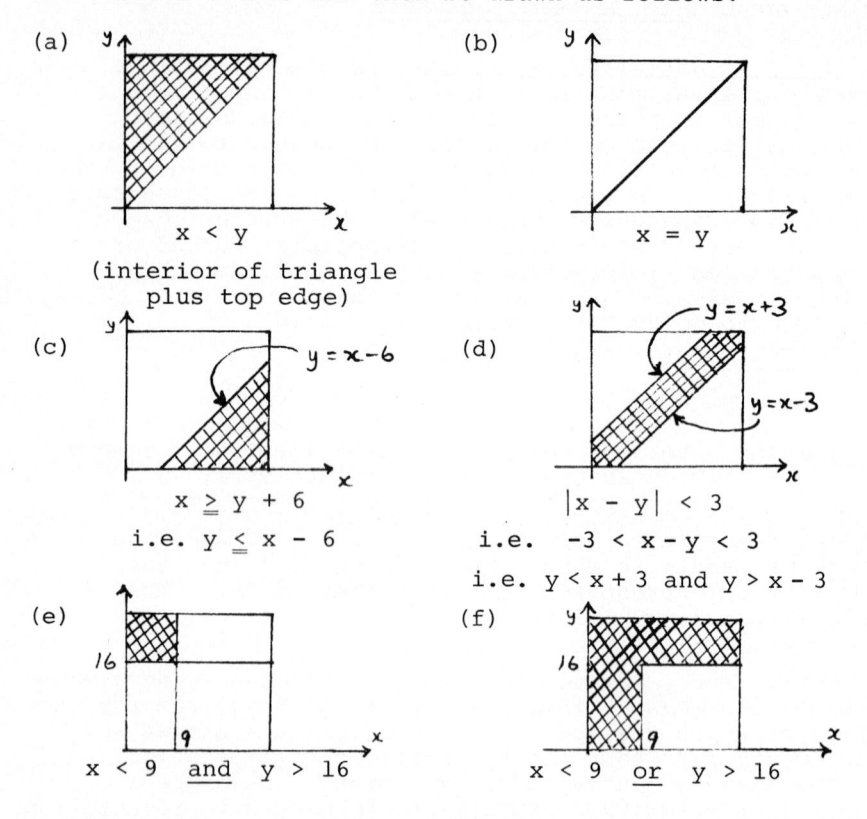

(a) x < y
(interior of triangle plus top edge)

(b) x = y

(c) x ≥ y + 6
i.e. y ≤ x - 6

(d) |x - y| < 3
i.e. -3 < x - y < 3
i.e. y < x + 3 and y > x - 3

(e) x < 9 <u>and</u> y > 16

(f) x < 9 <u>or</u> y > 16

<u>Notice</u>: (i) Taking the graphical representation of the sample space in the plane improves our idea of the situation.

(ii) The use of the words <u>and</u> and <u>or</u> in parts (e) and (f) signals the <u>intersection</u> and <u>union</u> of the two events {(x,y) : x < 9} and {(x,y) : y > 16}.

5. EVENTS HAPPENING TOGETHER

<u>EXAMPLE</u>. Explain how two events can happen together.

<u>Solution</u>. A person is chosen at random from the population of Edinburgh and sex and hair colour are noted. Define events as follows:

    M = person is a man,   W = person is a woman,

 F = person has fair hair, D = person has dark hair.

(We will take it that every person's hair can be classed as either dark or fair.)

We can think of a sample space for this experiment as a rectangle S representing all people in the town.

We divide the rectangle vertically into men and women and horizontally into people with dark and fair hair. Every person chosen lies somewhere in the rectangle. If the person chosen is a dark haired woman, she is represented by a point in the small rectangle in the top right hand corner. The choice of such a woman then means that the events D and W have both occurred simultaneously. In effect the small rectangle in the top right hand corner represents D ∩ W, i.e. the intersection of the events D and W.

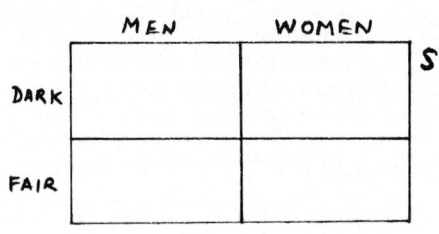

Notice: Some events cannot happen together. For example in the above example the events M and W cannot happen together, because their intersection is empty, i.e. M ∩ W = ∅. Such events are called <u>mutually exclusive</u> or <u>disjoint</u>. So two events or sets A and B are <u>mutually exclusive</u> or <u>disjoint</u> if A ∩ B = ∅. We can think of such sets as <u>not overlapping</u> in a Venn diagram.

EXAMPLES TO DO: Pages 7 - 8: Exs. 5, 6.

6. THE PROBABILITY OF AN EVENT

In the following example we introduce the idea of the probability of an event. Explanation together with the formal definition of probability follow below the example. For the moment you can think of the probability of an event as a measurement of the likelihood of that event on a scale from 0 to 1.

EXAMPLE. Refer to the experiment in the solution in §5, in which a person is chosen at random in Edinburgh and sex and hair colour are noted. Consider the representation of the sample space as a rectangle together with the events M, W, D and F. We now give figures attached to each of the subrectangles as shown. These figures represent the probabilities of each of the mutually exclusive events represented by the four sub-rectangles. Find the probabilities of each of the events M, W, D, F, M ∩ D and M ∪ D.

	MEN	WOMEN
DARK	0.32	0.24
FAIR	0.18	0.26

Solution.   P(M) = 0.32 + 0.18 = 0.50 .

P(W) = 0.24 + 0.26 = 0.50.   P(D) = 0.32 + 0.24 = 0.56.

P(F) = 0.18 + 0.26 = 0.44.   P(M ∩ D) = 0.32.

P(M ∪ D) = 0.32 + 0.24 + 0.18 = 0.74.

Notice: You can think of probability as sand - 1 gram in total attached to the experiment. For each event (i.e. each subset of the sample space) you can think of laying down a certain mass of sand from the 1 gram total to show how probable that event is. So in the above the probability of a fair haired man is 0.18, while the probability of a dark haired woman is 0.24.

The probability assigned to the union of two mutually exclusive events is the sum of their probabilities. So the probability that the person chosen is a man is 0.32 + 0.18 = 0.50.

Note also that the total probability across the whole sample space is 0.32 + 0.18 + 0.24 + 0.26 = 1.00, and this meets the demand that the total mass of "probability sand" over the whole sample space is 1 gram.

This example illustrates the basic requirements for the concept of the probability of an event. We can spell them out formally as follows.

Definition: Let S be the sample space for a random experiment. Then with every event A we associate a number called the probability of A meeting the following requirements:

(1) $0 \leq P(A) \leq 1$.

(2) $P(\emptyset) = 0$ and $P(S) = 1$.

(3) For every pair of mutually exclusive events A and B,
$$P(A \cup B) = P(A) + P(B).$$

(4) For every sequence of pairwise mutually exclusive events $\{A_n\}$,
$$P(A_1 \cup A_2 \cup A_3 \cup \ldots) = \sum_{i=1}^{\infty} P(A_i).$$

Notice: It is easy to deduce from property (3) that for a set of n events which are pairwise mutually exclusive

(3a) $$P(A_1 \cup A_2 \cup \ldots \cup A_n) = \sum_{i=1}^{n} P(A_i).$$

We do need property (4) however to cover the case of an infinite sequence of events.

7. THE PROBABILITY OF THE UNION OF TWO EVENTS

RESULT. For every pair of events A and B,
$$P(A \cup B) = P(A) + P(B) - P(A \cap B).$$

Proof. In the diagram let the probabilities associated with the mutually exclusive pieces be denoted by x, y and z as shown. Then using property (3a) from §6 with n = 3 gives

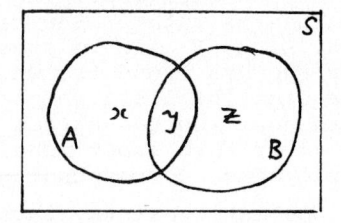

$P(A \cup B) = x + y + z$ while
$P(A) + P(B) - P(A \cap B) = (x+y) + (y+z) - y = x + y + z.$

Notice: (i) This important result applies to <u>every</u> pair of events A and B. Property (3) in §6 is just the special case of this result when $A \cap B = \emptyset$.

(ii) The result can be extended to three or more events. For example, for every three events A, B, C we have

$$P(A \cup B \cup C) = P(A) + P(B) + P(C)$$
$$- P(A \cap B) - P(B \cap C) - P(C \cap A)$$
$$+ P(A \cap B \cap C).$$

This can be proved from the above result using the distributive law

$$X \cap (Y \cup Z) = (X \cap Y) \cup (X \cap Z),$$

mentioned in §1. Extensions to four or more events follow a similar pattern.

EXAMPLES TO DO: Page 8: Exs. 7 - 12.

EXAMPLES 1

1. Let $S = \{1,2,3,4,5,6,7,8\}$, $A = \{1,2,3,4\}$, $B = \{3,4,5,6\}$, $C = \{1,3,5,7\}$. Find the sets

$\overline{A} \cap B$, $(A \cup B) \cap C$, $(A \cap B) \cup (B \cap C)$, $\overline{A} \cap \overline{B} \cap \overline{C}$, $(\overline{A \cup B}) \cap C$.

2. A coin is tossed until a head is obtained, and the number of attempts needed is noted. Describe a sample space for this experiment.

3. The president and vice president of a society are chosen from a pool of five people, namely A, B, C, D and E. Describe a sample space for this experiment.

4. A man invites six friends A, B, C, D, E, F to a party, and notes down whether they accept (Y) or refuse (N). Describe a sample space for this experiment.

5. A girl is trying to find two friends to go to the theatre with. She has a list of five friends J, K, L, M, N and telephones them in order stopping when she has found two who will go or when she has exhausted the list, whichever comes first. She then notes down who has agreed to go. Describe a sample space for this experiment. Give the subsets of the sample space corresponding to the following events:-

(a) She makes exactly three phone calls.
(b) She makes exactly five phone calls.
(c) Two consecutive phone calls produce people who agree to go with her.

6. A man and a woman (both over 100 cm and under 300 cm tall) are chosen at random and their heights, X cm for the man and Y cm for the woman are noted. Describe and sketch in the xy-plane a sample space for this experiment. Indicate on separate diagrams with shading the following events:-

(a) The heights of the man and the woman are equal.
(b) The woman is exactly 10 cm taller than the man.
(c) The man is more than 20 cm taller than the woman.
(d) Both the man and the woman are over 180 cm tall.
(e) The average height is 170 cm.
(f) The difference in heights is at least 10 cm.

7. Let A and B be events in sample space S with $P(A) = 0.4$, $P(B) = 0.3$ and $P(A \cup B) = 0.5$. Find $P(A \cap B)$, $P(\bar{A} \cap B)$ and $P(\bar{A} \cap \bar{B})$.

8. If $P(A) = 0.8$ and $P(B) = 0.5$, find the maximum and minimum possible values for $P(A \cap B)$.

9. Prove that a set with n members has 2^n subsets.

10. Use Venn diagrams to prove the result that for all sets A, B, C,

$$A \cap (B \cup C) = (A \cap B) \cup (A \cap C),$$

mentioned in §1(ii). Deduce that for all events A, B in a sample space S,

$$P(A) = P(A \cap B) + P(A \cap \bar{B}),$$

and draw a diagram to illustrate this.

11. Animals in a certain population may have one, both or none of two characteristics X and Y. It is known that the probability of having X alone is 0.15, the probability of having Y alone is 0.35 and the probability of having neither X nor Y is 0.30. Find the probability of having both X and Y.

12. Derive the formula for $P(A \cup B \cup C)$ mentioned in §7(ii) using the method suggested there.

II. COUNTING CASES, EQUALLY LIKELY OUTCOMES THE HYPERGEOMETRIC DISTRIBUTION

8. THE MULTIPLICATION PRINCIPLE

EXAMPLE. The menu in a restaurant offers 4 soups, 7 meat courses and 5 desserts. How many different three course meals can be made up?

Solution. Answer = $4 \times 7 \times 5 = 140$.

Notice: This is an important idea. Effectively you can take 4 routes through the soups; each of these routes then divides into 7 ways through the meat courses. So there are effectively 28 different routes through the first two courses. Each of these 28 routes can then divide into 5 routes through the desserts to give a total of 140 routes altogether.

9. PERMUTATIONS

Definition. A permutation of n objects is an arrangement of the objects in some order.

We illustrate with the following example.

EXAMPLE. Suppose that we have seven people X, Y, Z, T, U, V, W, who are to be seated in a row of chairs facing us. Here are two questions.

(a) How many arrangements are possible if there are seven chairs in the row?

(b) How many arrangements are possible if there are only four chairs in the row?

Solution. (a) There are seven seats available. Think first of filling the first seat on the left. There are 7 ways of doing this. Now fill the second seat from the left. There are 6 ways of doing this (i.e. using the 6 people still available). This makes 42 ways of filling the first two seats. There are then 5 ways of filling the third seat and so on until there is only 1 way of filling the last seat (i.e. only one person remaining). So the total number of ways of filling the row of 7 seats is $7 \times 6 \times 5 \times 4 \times 3 \times 2 \times 1 = 7! = 5040$.

(b) Here there are only four seats available, so that there will be 3 disappointed people left out. Proceeding as in (a) we find there are $7 \times 6 \times 5 \times 4$ ways of filling the 4 seats, i.e. 840 ways. [We can write it as $7!/3!$ if we like.]

Notice: The notation $n! = n(n-1)(n-2)\ldots3.2.1$ is standard and is read as n factorial. So $1! = 1$,

2! = 2, 3! = 6, 4! = 24, 5! = 120, 6! = 720.   It is a
convention (and you will need it) that 0! = 1.

Generalising the reasoning in the above example shows that (a) there are n! ways of filling n spaces with n different objects, and (b) there are n!/(n-k)! ways of filling k spaces where there are n different objects available (with n ≥ k).

10. AN EXAMPLE ON PERMUTATIONS

EXAMPLE. Six people X, Y, Z, T, U, V are placed at random on a row of six seats. How many arrangements are possible altogether? Find also the number of arrangements in which

    (a) X occupies the left hand end seat,
    (b) X and Y occupy the end seats,
    (c) X and Y sit side by side.

Solution. As in §9, the total number of arrangements is 6! = 720. For the other parts proceed as follows.

(a)   X _ _ _ _ _

Imagine X on the left hand end seat. We then have a free choice to fill the other 5 seats. We can do this in 5! (i.e. 120) ways. So the answer is 120.

(b) Consider the two cases

        X _ _ _ _ Y   and   Y _ _ _ _ X.

In each of these cases there are 4! (i.e. 24) ways of filling the intermediate seats. So the answer required is 24 + 24 = 48.

(c) Consider 10 cases, namely the five given by

      XY....   .XY...   ..XY..   ...XY.   ....XY

and a similar five with X and Y reversed. In each of these cases there are 4! (i.e. 24) arrangements. So the total number of arrangements possible is 240.

Notice: In this problem the order in which the people are arranged on the seats is important in the sense that (for example) in (b) the arrangements

              XZTUVY   and   YZTUVX

are regarded as different. This importance of order is a feature of permuations. Contrast the situation for combinations in §11 - 12.

EXAMPLES TO DO: Page 16: Exs. 1, 3, 5, 6.

11. COMBINATIONS AND BINOMIAL COEFFICIENT NOTATION

EXAMPLE. Six people are at a party where the milk runs out. Two people are to be chosen to go for milk. How many different choices are possible?

Solution. Let the people be A, B, C, D, E, F. The possible pairs are then

AB,AC,AD,AE,AF,BC,BD,BE,BF,CD,CE,CF,DE,DF,EF,

i.e. 15 possible pairs.

Notice: (i) Here order is irrelevant in the sense that the pair AB is the same as the pair BA.

(ii) A combination of k objects out of a total of n objects is a selection or choice of k objects out of the n, order within the group being irrelevant. So the example above shows that the number of combinations of 2 objects out of 6 is 15. We have a notation for this, namely $\binom{6}{2} = 15$, which is explained below. The old notation for this $_6C_2$ (read as 6 choose 2) still has much to commend it though it has fallen from favour at present.

(iii) The following result gives a method of calculating $\binom{n}{k}$ in general, without counting cases as in the above example.

RESULT. The number of combinations (i.e. selections) of k objects from a group of n different objects is

$$\binom{n}{k} \quad \text{where} \quad \binom{n}{k} = \frac{n!}{(n-k)!\,k!} \,.$$

Proof. Take a set of n objects. We now perform a two stage process on these n objects, as follows. As the first stage, choose k objects out of the n objects: there will be Q such selections where Q is the number $\binom{n}{k}$ which we are seeking. Then, as the second stage, arrange the k objects chosen in a row: there are k! such arrangements for each selection. So, by the multiplication principle, the number of ways of performing the two stage process is $Q \times k!$.

However, if we look at the two stage process as a whole, it really just fills a row of k seats where n people are available, and we know from §9 that the number of ways of doing this is $n!/(n-k)!$. So we conclude that $Q.k! = \dfrac{n!}{(n-k)!}$ so that $Q = \dfrac{n!}{(n-k)!\,k!}$ as required.

Notice: (i) For a positive integer n and an integer k with $0 \leq k \leq n$, we define the <u>binomial coefficient</u> $\binom{n}{k}$ by $\binom{n}{k} = \frac{n!}{(n-k)!k!}$. In general, rather than working with all the factorials, it is easier to use the equivalent formula $\binom{n}{k} = \frac{n(n-1)(n-2)\ldots(n-k+1)}{k!}$.

For example, $\binom{12}{3} = \frac{12 \cdot 11 \cdot 10}{3!} = 220$.

(ii) The following easily checked facts about binomial coefficients have to be known:

$$\binom{n}{0} = 1, \quad \binom{n}{1} = n, \quad \binom{n}{n-k} = \binom{n}{k},$$

e.g. $\binom{5}{0} = 1, \quad \binom{5}{1} = 5, \quad \binom{10}{7} = \binom{10}{3}$.

12. AN EXAMPLE ON COMBINATIONS

<u>EXAMPLE.</u> A ship's company of 15 men consists of 5 officers (one of whom is the captain) and 10 ratings. Three men are to be chosen to visit a sick children's hospital. How many groups of three are possible altogether? Find also the number of groups that can be made up if

    (a) the captain must be in the group,
    (b) the group consists of 1 officer and 2 ratings,
    (c) the group must contain at least 1 rating.

<u>Solution.</u> The number of possible groups is $\binom{15}{3} = 455$. For the other parts proceed as follows.

(a) If the captain is to be included we really want the number of ways of choosing two partners, i.e. $\binom{14}{2} = 91$. So the answer is 91.

(b) Number of ways of choosing the officer = $\binom{5}{1} = 5$.

    Number of ways of choosing the ratings = $\binom{10}{2} = 45$.

So, by the multiplication principle, the total number of possible groups is $5 \times 45 = 225$.

(c) The number of groups in which <u>no</u> rating is included, i.e. all officers, is $\binom{5}{3} = 10$.

So, by subtraction from the overall total, we see that the number of groups with at least one rating is given by $455 - 10 = 445$.

Notice: In the above example <u>order is unimportant</u> in the sense that what matters is the <u>set of people chosen not</u> the order in which they are chosen. As already noted in §11, this unimportance of order is a feature of problems involving <u>combinations</u>.

EXAMPLES TO DO: Page 16: Exs. 2, 4, 7.

13. EQUALLY LIKELY OUTCOMES

In some random experiments all outcomes are equally likely. For example if you look at the ship's company in §12 sending the deputation to the sick children's hospital, then if the three names are drawn at random from a hat it follows that all possible groups of three are equally likely. The following result applies in such a situation.

<u>RESULT</u>. Let E be a random experiment in which <u>all outcomes are equally likely</u>. Let A be an event. Then

$$P(A) = \frac{\text{Number of outcomes favourable to } A}{\text{Total number of outcomes of } E}.$$

<u>Notice</u>: (i) Effectively the result tells you to take the fraction given by the number of outcomes favourable to the event divided by the total number of outcomes.

(ii) The result is quite useful but do notice the proviso that all outcomes must be equally likely.

(iii) The following example together with those in §14 and §15 illustrate the use of this result.

<u>EXAMPLE</u>. (Continuation of the example in §12) A ship's company of 15 men consists of 5 officers (one of whom is the captain) and 10 ratings. Three men are chosen at random to visit a sick children's hospital. Find the probability that

(a) the group consists of 2 officers and 1 rating,
(b) the group consists of the captain, one other officer and one rating.

<u>Solution</u>. Here all outcomes are equally likely (i.e. all groups of three are equally probable). So we can use the above result.

(a) $P(2O \ \& \ 1R) = \dfrac{\text{No. of ways of choosing } 2O \ \& \ 1R}{\text{Total no. of ways of choosing 3}}$

$= \dfrac{\binom{5}{2}\binom{10}{1}}{\binom{15}{3}} = \dfrac{10.10}{455} = \dfrac{100}{455} = \dfrac{20}{91}.$

(b) $P(C + \text{other } O + R) = \dfrac{1 \cdot \binom{4}{1}\binom{10}{1}}{\binom{15}{3}} = \dfrac{40}{455} = \dfrac{8}{91}$.

Notice: The multiplication principle is used in this example to count cases in terms like

$$\binom{5}{2}\binom{10}{1} \text{ in (a) and } 1 \cdot \binom{4}{1}\binom{10}{1} \text{ in (b).}$$

14. THE HYPERGEOMETRIC DISTRIBUTION

The method used in the above example (i.e. §13) illustrates a standard type of situation in which the distribution of probability is said to be a <u>hypergeometric distribution</u>. The following result gives the basic formula.

RESULT. Suppose that we have a finite set of N items, made up of n_1 items of the first type and n_2 items of the second type, (so that $N = n_1 + n_2$). Suppose now that a subset of K items is drawn at random from the original set, (all possible subsets of K items being equally likely). Then the probability that the subset consists of k_1 items of the first type and k_2 items of the second type is

$$\dfrac{\binom{n_1}{k_1}\binom{n_2}{k_2}}{\binom{N}{K}} .$$

Notice: (i) In the formula, $N = n_1 + n_2$, $K = k_1 + k_2$.

(ii) The formula can clearly be extended if there are more than two types of item, e.g. look at §15(a) and (b).

EXAMPLE. A factory contains 12 machines of which 8 are safe and 4 are dangerous. A factory inspector chooses 3 machines at random for inspection. Find the probability that he chooses

(a) exactly two dangerous machines,
(b) at most one dangerous machine,
(c) at least one dangerous machine.

Solution. This is a <u>hypergeometric</u> situation. Let X denote the number of dangerous machines found. Use the letters S for safe, and D for dangerous. We apply the above formula as follows.

(a) $P(X = 2) = P(2D, 1S) = \dfrac{\binom{4}{2} \cdot \binom{8}{1}}{\binom{12}{3}} = \dfrac{6 \cdot 8}{220} = \dfrac{48}{220} = \dfrac{12}{55}$.

(b) $P(X \leq 1) = P(X=0) + P(X=1) = P(0D, 3S) + P(1D, 2S)$

$$= \frac{\binom{4}{0} \cdot \binom{8}{3}}{\binom{12}{3}} + \frac{\binom{4}{1} \cdot \binom{8}{2}}{\binom{12}{3}} = \frac{56}{220} + \frac{112}{220} = \frac{42}{55}.$$

(c) $P(X \geq 1) = 1 - P(X=0) = 1 - \frac{56}{220}$ (from part (b))

$$= \frac{164}{220} = \frac{41}{55}.$$

Notice: The method illustrated in part (c) in which the probability of the complementary event (i.e. $X=0$) is subtracted from 1 is a technique well worth remembering for other problems.

15. EXAMPLE. From an ordinary pack of 52 cards, we draw 4 cards at random without replacement†. Find the probability that the 4 cards consist of

(a) 1 diamond, 1 heart and 2 black cards,
(b) 2 aces and 2 face cards (i.e. King, Queen or Jack).

Solution. We use an extended version of the formula in the result in §14, in which extra factors are included in the top line to correspond to more than two types of item.

(a) $P(1D, 1H, 2\ Black) = \dfrac{\binom{13}{1} \cdot \binom{13}{1} \cdot \binom{26}{2}}{\binom{52}{4}} = 0.2029$.

(b) $P(2\ aces, 2\ face) = \dfrac{\binom{4}{2} \cdot \binom{12}{2} \cdot \binom{36}{0}}{\binom{52}{4}} = 0.001463$.

† Remark. (Replacement) If from four articles A, B, C and D we draw two articles without replacement then the possible pairs are

AB, AC, AD, BC, BD, CD.

If however we draw the two articles with replacement the first article is replaced in the pool before the second article is drawn. So in this case the pairs AA, BB, CC and DD are also possible.

EXAMPLES TO DO: Pages 16 - 17: Exs. 8 - 16.

EXAMPLES 2

1. A doctor visits 5 houses every day. In how many different orders can he make the visits?

2. A company has 24 employees. Every day two of them are chosen to go to the bank. In how many different ways can the choice be made?

3. A society has 20 members. Three people are required to fill the posts of president, secretary and treasurer. In how many ways can the posts be filled?

4. An airline has 6 pilots, 5 co-pilots and 12 stewardesses. In how many ways can a crew of four (i.e. 1 pilot, 1 co-pilot and 2 stewardesses) be made up?

5. At a society dinner, guests sit on one side of the top table in a row of seven seats. The seven seats are to be filled by the president, the speaker, a social climber and four other people. Find the total number of seating arrangements possible. In how many of these arrangements do we find that

   (a) the social climber sits on the middle seat,
   (b) the speaker sits on the middle seat and the social climber sits beside him and on his right,
   (c) the social climber sits with the speaker as one neighbour and the president as the other,
   (d) the social climber sits beside the speaker?

6. Code words are to be made up of 6 letters A and 4 letters B (e.g. AAABBBAABA). How many possible words are there? How many words start and end with the letter A ? How many start with A and end with B ?

7. A ship's crew consists of 4 Americans, 2 Germans and 5 Norwegians. A set of four people is to be chosen to row ashore in an open boat. How many different sets of four are possible? How many sets include exactly 2 Norwegians? How many sets have at least one person of each nationality?

8. A set of 12 locomotives consists of 7 fitted with snowplough and 5 without. Every night 3 locomotives (chosen at random) are kept in a certain town. Find the probability that on any given night these 3 locomotives consist of

   (a) 2 with snowplough and 1 without,
   (b) at least one with snowplough.

9. A lifeboat contains 6 men and 4 women. Four people to row are chosen at random. Find the probability that the four people consist of

   (a) 2 men and 2 women, (b) at most 1 woman,
          (c) at least 1 woman.

10. An airline has 6 planes of type A and 6 of type B. Their day to day distribution is random. One night a fire in a hangar destroys 4 of the planes. Find the probability that

    (a) 3 planes of type A and 1 of type B are lost,
    (b) all 4 planes lost are of the same type.

11. A box contains twelve cards numbered from 1 to 12. Two cards are drawn without replacement and the numbers on them are noted. Find the probability that

    (a) the sum of the numbers is exactly 22,
    (b) the sum of the numbers is at least 20,
    (c) the lower of the two numbers is 5.

12. From an ordinary pack of 52 cards, a hand of 13 cards is dealt. Find the probability that the hand contains n aces for $n = 0, 1, 2, 3, 4$, giving the answers to 3 decimal places. [You will need a calculator.] Explain why the five answers add up to 1.

13. From an ordinary pack of 52 cards, a hand of 13 cards is dealt. Find the probability that

    (a) all 13 cards are red,
    (b) all 13 cards are 10 or below (ace counts high).

14. Eight people - 4 men and 4 women - sit round a circular table. Miss Y wishes not to sit beside Mr. X. Find the probability that her wish will be granted in the cases when

    (a) the seating is entirely random,
    (b) men and women alternate round the table.

15. Among a group of 7 paintings there are 5 genuine and 2 fake. Two art experts are called to examine them. Each then has to declare (independently of the other) which two paintings he thinks are fake. If the experts just guess at random, what is the probability that they will both identify the same two paintings as fake?

16. Answer (a) and (b) in Ex.11 above if the two cards are drawn with replacement.

III. CONDITIONAL PROBABILITY

16. In many situations, once more information becomes available we revise our estimates for the probability of further events happening. For example, as you climb out of a hotel bed in the morning you may estimate the probability of receiving breakfast as 0.99. However if the fire alarm then sounds and you find yourself standing in the street watching the firemen arriving you may revise your estimate of the probability of receiving breakfast to 0.60.

This is the idea of <u>conditional probability</u>. The notation $P(B|A)$ denotes the probability of B happening given that A has happened. We read $P(B|A)$ as the <u>probability of B given A</u>. So, in the above example, as you climb out of bed

$$P(\text{Breakfast}) = 0.99$$

but once the alarm goes off and you are in the street

$$P(\text{Breakfast}|\text{Alarm}) = 0.60.$$

17. <u>EXAMPLE</u>. 6 men and 5 women have dark and fair hair according to the scheme:

    <u>Men</u>:   D D D D F

    <u>Women</u>: D D F F F

A person is chosen at random and hair colour and sex are noted. Let A and B be the events given by

    A = Person chosen is fair,
    B = Person chosen is a woman.

Find $P(A)$, $P(B)$, $P(A|B)$, $P(B|A)$ and $P(A \cap B)$.

<u>Solution</u>. $P(A) = \frac{4}{11}$ and $P(B) = \frac{5}{11}$.

$$P(B|A) = \frac{3}{4} \quad \text{and} \quad P(A|B) = \frac{3}{5}.$$

$$\text{Also} \quad P(A \cap B) = \frac{3}{11}.$$

<u>Notice</u>: (i) In this example we have that

$$P(A \cap B) = P(B|A)P(A) = \frac{3}{11},$$

and also

$$P(A \cap B) = P(A|B)P(B) = \frac{3}{11}.$$

This is not just chance. It illustrates a general rule, which we shall use in various forms later. §18 gives a statement of the rule while §19 - 21 give examples of how to use it.

18. CONDITIONAL PROBABILITY: THE BASIC FORMULA

<u>RULE</u>. For every pair of events A and B,

$$P(A \cap B) = P(A|B) P(B) = P(B|A) P(A).$$

<u>Notice</u>: (i) §17 illustrates the rule.

(ii) The rule is often useful in the form

$$P(B|A) = \frac{P(A \cap B)}{P(A)}$$

and in fact in this form it can be used as a <u>definition</u> of the <u>conditional probability</u> of the event $\overline{B}$ occuring given that A has occurred.

19. EXAMPLE.
A box contains 7 red and 2 white balls. Two balls are drawn out without replacement. Find the probability that

(i) the first ball is red and the second ball is white,

(ii) both balls are red.

<u>Solution</u>. (i) Take events A and B given by

A = the first ball is red,
B = the second ball is white.

We want $P(A \cap B)$. So using the formula from §18 gives

$$P(A \cap B) = P(B|A) P(A) = \frac{2}{8} \cdot \frac{7}{9} = \frac{14}{72} = \frac{7}{36}.$$

(ii) Take events C and D given by

C = the first ball is red,
D = the second ball is red.

We want $P(C \cap D)$. So again from §18 we have

$$P(C \cap D) = P(D|C) P(C) = \frac{6}{8} \cdot \frac{7}{9} = \frac{42}{72} = \frac{7}{12}.$$

<u>Notice</u>: In calculating $P(A \cap B)$ we had to choose between $P(A|B) P(B)$ and $P(B|A) P(A)$. We chose the latter because A happens <u>first</u>. Once we know a red ball has been drawn on the <u>first</u> choice it is then clear that there are 6 red and 2 white remaining so that $P(B|A) = 2/8$.

On the other hand, finding $P(A|B)$ demands the probability of an earlier event given that a later event has occurred, and this is not so easy. We actually address such problems in connection with Bayes' Theorem in §23.

<u>EXAMPLES TO DO</u>:   Pages 24 - 26: Exs. 1, 2, 3, 15.

20. EXAMPLE. A coin is tossed three times. Find the probability of three heads.

Solution. Let H_i be the event of tossing a head on the ith toss ($i = 1, 2, 3$). We want $P(H_1 \cap H_2 \cap H_3)$.

So $P(H_1 \cap H_2 \cap H_3) = P(H_2 \cap H_3 \mid H_1) P(H_1)$ (by §18)

$\qquad\qquad\qquad = P(H_2 \cap H_3) P(H_1)$

$\qquad\qquad\qquad\qquad$ (since the first toss does not affect the others)

$\qquad\qquad\qquad = P(H_3 \mid H_2) P(H_2) P(H_1)$ (by §18)

$\qquad\qquad\qquad = P(H_3) P(H_2) P(H_1)$

$\qquad\qquad\qquad\qquad$ (since the second toss does not affect the third)

$\qquad\qquad\qquad = \tfrac{1}{2} \cdot \tfrac{1}{2} \cdot \tfrac{1}{2} = \tfrac{1}{8}$.

Notice: (i) Here we have spelled out the details to show how this is covered by the rule of §18. In practice you can write the answer down, particularly after looking at the idea of independence in §26.

(ii) For more insight into the situation think of 8 people each tossing a coin attempting to toss 3 heads in a row. After the first toss, there will be roughly 4 survivors (because roughly 4 people will have tossed a tail and so have been eliminated). After the second toss there will similarly be roughly 2 survivors and after the third toss there will be roughly 1 survivor from the original 8 candidates. So it seems reasonable that the probability of tossing 3 heads in a row is 1/8.

EXAMPLES TO DO: Pages 24 - 25: Exs. 4, 14.

21. EXAMPLE. (A standard type) A lot contains 12 articles of which 8 are good and 4 are defective. The articles are inspected one after another in random order. Find the probability that the ninth article inspected is the fourth defective article found.

Solution. Define events A and B by

A = the first eight articles are 5 good, 3 defective,
B = the ninth article is defective.

We then want $P(A \cap B)$, for which we use §18. So

$$P(A \cap B) = P(B \mid A) P(A) = \frac{1}{4} \cdot \frac{\binom{8}{5}\binom{4}{3}}{\binom{12}{8}} = \frac{56}{495}.$$

Notice: (i) In calculating P(A) we are just making a direct application of the formula for the hypergeometric distribution given in the result in §14. Be on the lookout for other similar situations.

(ii) In finding $P(A \cap B)$ we had to choose here between using $P(A|B) P(B)$ and $P(B|A) P(A)$ from §18. We chose the latter because $P(B|A)$ is much easier to find than $P(A|B)$: this is because <u>A occurs first</u>.

(iii) An attempt to do the problem with an event like "the first 9 articles are 5 good and 4 defective" is invalid and gives the wrong answer because it allows the fourth defective article to be found <u>before</u> the the ninth choice.

EXAMPLES TO DO: Page 24: Exs. 5, 6.

22. PARTITIONS OF THE SAMPLE SPACE

A partition of the sample space effectively carves the sample space into a finite number of non-overlapping pieces. Here is the definition.

<u>Definition</u>. Events $A_1, A_2, \ldots, A_n$ form a partition of the sample space S if the following conditions hold:

(i) $A_i \cap A_j = \emptyset$ $(i \neq j)$;

(ii) $A_1 \cup A_2 \cup \ldots \cup A_n = S$;

(iii) $P(A_i) > 0$ for every i.

<u>EXAMPLE</u>. On a pig farm there are three types of pig, namely X, Y and Z. 20% are of type X, 50% are of type Y and 30% are of type Z. A certain disease affects 8% of type X, 9% of type Y and 4% of type Z. Find the probability that a randomly chosen pig on the farm is affected by the disease.

<u>Solution</u>. The set S of all pigs on the farm is partitioned into 3 subsets X, Y, Z and the subset of diseased pigs D straddles all three subsets of the partition as shown in the diagram. It is then clear that

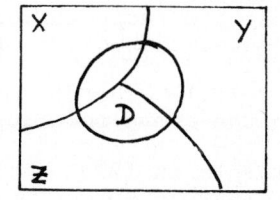

$$P(D) = P(D \cap X) + P(D \cap Y) + P(D \cap Z)$$
$$= P(D|X) P(X) + P(D|Y) P(Y) + P(D|Z) P(Z)$$
$$= \frac{8}{100} \cdot \frac{20}{100} + \frac{9}{100} \cdot \frac{50}{100} + \frac{4}{100} \cdot \frac{30}{100} = 0.073.$$

EXAMPLES TO DO: Page 24: Ex. 7.

23. BAYES' TEOREM

EXAMPLE. Three factories A, B, C produce 50%, 40% and 10% respectively of the radios of a certain make. Of their outputs 2%, 7% and 3% respectively are defective. If a radio chosen at random in the company store is found to be defective, find the probability that it came from factory A.

<u>Solution</u>. Let A, B, C be the events that a radio comes from factories A, B, C. Let D be the event that a radio is defective. We want $P(A|D)$. Use the formula

$$P(A|D) P(D) = P(D|A) P(A) \quad \text{from §18.}$$

So $P(A|D) = \dfrac{P(D|A) P(A)}{P(D)} = \dfrac{P(D|A) P(A)}{P(D \cap A) + P(D \cap B) + P(D \cap C)}$

$= \dfrac{P(D|A) P(A)}{P(D|A) P(A) + P(D|B) P(B) + P(D|C) P(C)}$

$= \dfrac{\frac{2}{100} \cdot \frac{50}{100}}{\frac{2}{100} \cdot \frac{50}{100} + \frac{7}{100} \cdot \frac{40}{100} + \frac{3}{100} \cdot \frac{10}{100}} = \dfrac{10}{41}.$

(Similarly we can show that $P(B|D) = \dfrac{28}{41}$ and $P(C|D) = \dfrac{3}{41}$.)

<u>Notice</u>: (i) This problem illustrates the use of Bayes' Theorem, which we state in (iv) below. Effectively Bayes' Theorem attempts to go backwards in the sense that it looks at what happens later and then gives the conditional probability that an earlier event happened.

(ii) The two vital features of problems on Bayes' Theorem are (a) ancestry, and (b) evidence. Within the problem there are generally two or more possible types of <u>ancestry</u> - here it is a case of originating in one of three different factories. And then there is <u>evidence</u> of something that happened later - here it is that the radio turned out to be defective. Bayes' Theorem then gives P(one type of ancestry|evidence).

(iii) The denominator P(D) in the above solution is found in the same way as P(D) in the example in §22. Effectively the sample space is carved into three parts corresponding to the different ancestries A, B, C and the event D straddles them as the diagram illustrates.

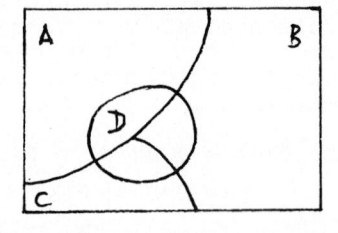

(iv) In such problems it is generally convenient to start (as in the above solution) from the formula

$$P(A|D)\ P(D) = P(D|A)\ P(A)$$

and then to proceed from there. However we can state Bayes' Theorem formally as follows:

<u>RESULT</u>. (<u>Bayes' Theorem</u>) Let $A_1, A_2, \ldots, A_n$ be a partition of sample space S, and let E be an event. Then for each $i = 1, 2, \ldots, n$,

$$P(A_i|E) = \frac{P(E|A_i)\ P(A_i)}{\sum_{k=1}^{n} P(E|A_k)\ P(A_k)}.$$

[In this result the denominator is just $P(E)$. Also the events A_i are the ancestries while E is the evidence.]

24. <u>EXAMPLE</u>. (<u>Also on Bayes' Theorem</u>) A bag contains seven coins of which two are double headed and five are normal. A coin is drawn from the bag and is tossed four times coming up heads each time. Find then the probability that this is a normal coin.

<u>Solution</u>. Take the events N = normal coin chosen,
  D = double headed coin chosen,
  $4H$ = four heads are tossed.

We want $P(N|4H)$. So start with

$$P(N|4H)\ P(4H) = P(4H|N)\ P(N) \quad \text{(from §18)}.$$

So $P(N|4H) = \dfrac{P(4H|N)\ P(N)}{P(4H)}$

$= \dfrac{P(4H|N)\ P(N)}{P(4H|N)\ P(N) + P(4H|D)\ P(D)}$

$= \dfrac{\left(\frac{1}{2}\right)^4 \cdot \frac{5}{7}}{\left(\frac{1}{2}\right)^4 \cdot \frac{5}{7} + 1 \cdot \frac{2}{7}} = \dfrac{5}{37}.$

<u>Notice</u>: (i) This follows the same pattern as the example in §23. Here there are <u>two</u> types of ancestry: the coin chosen is either double headed or normal. The evidence is the fact that four heads are tossed.

(ii) The probability of tossing 4 heads in 4 tosses of a normal coin is $(\tfrac{1}{2})^4 = 1/16$. See §20 if you are unhappy about this.

<u>EXAMPLES TO DO</u>:   Pages 24 - 25: Exs. 8 - 13.

EXAMPLES 3

1. A pool of seven men and six women come from Glasgow and Edinburgh as follows:

        MMMWW   MMMMWWWW

        GLASGOW   EDINBURGH

A person is chosen at random from the pool. Let E, G, M, W denote the events that the person is from Edinburgh, from Glasgow, a man, a woman respectively. Find $P(M)$, $P(W)$, $P(G)$, $P(G|W)$, $P(W|G)$, $P(G|M)$, $P(E|M)$, $P(G \cap W)$.

2. A box contains 4 red, 5 white and 2 blue balls. Two balls are drawn out one after the other without replacement. Find the probability that

    (a) the first ball is red and the second is white,
    (b) the first ball is red and the second is blue,
    (c) both balls are red.

3. Two cards are drawn from an ordinary pack without replacement. Find the probability that

    (a) both cards are red,
    (b) the first card is red and the second is black,
    (c) one card is red and the other is black.

4. A dice is thrown three times. Find the probability of

    (a) three sixes,
    (b) a six, then an odd number, then another six,
    (c) three different numbers (e.g. 5, 4, 1),
    (d) three numbers the same (e.g. 5, 5, 5).

5. A set of 10 articles consists of 8 good ones and 2 defectives. The articles are tested one after another in random order. Find the probability that

    (a) the first four tests produce good articles,
    (b) the second defective article is found on the fourth test.

6. A hand of 13 playing cards consists of 4 aces and 9 others. The cards are laid face down on a table and are turned over one by one until all the aces have been found. Find the probability that the fourth ace is found when the tenth card is turned over.

7. A box contains 6 yellow, 2 grey and 3 blue balls. A ball is drawn out and is thrown away. A second ball is then drawn. Find the probability that this second ball is yellow.

8. Shirts are supplied to a shop by three suppliers A, B and C, who provide 30%, 50% and 20% of all the

shirts respectively. On average 40%, 15% and 5% of the shirts supplied by A, B, C are found to be defective. A shirt chosen at random in the shop is found to be defective. Find the probability that this shirt came from supplier A.

9. A company has three factories K, L, M all making carpet tiles, their outputs being respectively 20%, 50% and 30% of the total output. The percentages of defective tiles among the outputs of A, B and C are 6%, 2% and 5% respectively. A tile chosen at random from stock is found to be defective. Find the probability that it came from factory K.

10. A bag contains 6 coins of which 5 are normal and 1 is double-headed. A coin is drawn from the bag and is tossed 3 times. It comes up heads each time. Find the probability that this is the double-headed coin.

11. A bag contains 2 coins of which one is normal and the other is double-headed. A coin is drawn from the bag and is tossed four times. It comes up heads each time. What is the probability that this is the double-headed coin? The other coin is then drawn from the bag and is tossed twice, coming up heads each time. What is then the probability that the first coin drawn is double-headed?

12. Boxes X and Y contain red and white balls. Box X contains 5 red and 7 white balls while box Y contains 4 red and 9 white balls. A ball is chosen at random from box X and is put into box Y. A ball is then chosen from box Y and its colour is noted. Find the probability that this ball is red.
It is now given that the ball drawn from box Y has turned out to be red. In the light of this find the probability that the ball transferred from X to Y was white. Find also the probability that another ball now drawn from box X will be white.

13. Two gunners A and B fire a gun at a target on a firing range. The probability that any shot will hit the target is h (for a shot from A) and is k (for a shot from B) and these probabilities are constant from shot to shot. An observer at a distance sees one of the gunners fire three shots all of which hit the target. Find the probability in terms of h and k that this gunner is A. The observer then sees the other gunner take over the firing of the gun and fire three shots all of which miss the target. Find now the probability that the first gunner was A. Find your answers explicitly in the case when $h = \frac{3}{4}$ and $k = \frac{1}{4}$.

14. A dice is thrown again and again. Find the probability that
(a) no six has been thrown after n throws,
(b) the first six is thrown on the nth throw.

15. A railway photographer knows that a certain train will leave a station pulled by two locomotives chosen at random from a pool of 16 locomotives of which 6 are Bo-Bos numbered B1, B2, ... , B6 and 10 are Co-Cos numbered C1, C2, ... , C10. He arrives late at the station and finds the train has just gone. He then asks a woman on the platform if she can tell him which locomotives were pulling the train. Here are three possible replies:

   (a) "Sorry, I know nothing about locomotives."
   (b) "One locomotive was a Bo-Bo but I know nothing about the other."
   (c) "One locomotive was the Bo-Bo numbered B5 but I know nothing about the other."

In the case of each reply find the probability that it was two Bo-Bo locomotives pulling the train.

16. The time is 2245 hours. The place is a television studio in London. The programme is a live national news and discussion programme. You are being interviewed on a one to one basis. The question is: "Briefly (say in a couple of sentences) explain Bayes' Theorem to the viewers in general terms."
Write down your two sentences carefully.

IV. INDEPENDENCE

25. In a search for a mathematical definition of the independence of two events, look at the following example.

EXAMPLE. A person X is chosen at random from the population of London and sex, height and eye colour are noted. Define events A, B, C by

   A = X is a man,
   B = X has blue eyes,
   C = X is over 6 feet tall (i.e. 183 cm).

Assign reasonable values to $P(A)$, $P(A|B)$ and $P(A|C)$ and discuss whether A and B are independent and whether A and C are independent.

Solution. It is fairly clear that the events A and B are independent of each other: being told that a person has blue eyes gives no information in general about their sex. On the other hand it is fairly clear that the events A and C are related (i.e. not independent) because if you are told a person is over 6 feet tall, it is then quite likely the person is a man.

We might reasonably assign the probabilities as

$$P(A) = 0.5, \quad P(A|B) = 0.5, \quad P(A|C) = 0.9.$$

Notice: This example suggests that if events A and B are independent then $P(A|B) = P(A)$. This is actually in accordance with the definition of independence given in §26, though the definition comes in a somewhat different form.

26. INDEPENDENCE OF TWO EVENTS

Definition. Two events A and B are <u>independent</u> if and only if

$$P(A \cap B) = P(A) P(B).$$

In some problems it is clear that two events are independent, in others you are told they are independent. To connect this definition with the idea of independence indicated in §25, look at the following result.

RESULT. Let A and B be events with $P(A) \neq 0$ and $P(B) \neq 0$. If A and B are independent then

   (i) $P(A|B) = P(A)$, and (ii) $P(B|A) = P(B)$.

Conversely, if either of (i) or (ii) holds then the events A and B are independent.

Proof. First suppose that A and B are independent.

So, by definition, $P(A \cap B) = P(A) P(B)$.
But, from the rule in §18,
$$P(A \cap B) = P(A|B) P(B).$$
So $P(A|B) P(B) = P(A) P(B)$,
from which we conclude that $P(A|B) = P(A)$, on cancelling the non-zero factor $P(B)$. Similarly we can prove that $P(B|A) = P(B)$.

For the converse, suppose that $P(A|B) = P(A)$. Then from the rule in §18,
$$P(A \cap B) = P(A|B) P(A) = P(A) P(B).$$
So A and B are independent.

<u>Notice</u>: The definition of the independence of two events A and B in the form
$$P(A \cap B) = P(A) P(B)$$
has the advantage of symmetry over the forms based on $P(A|B) = P(A)$.

27. <u>EXAMPLE</u>. A coin is tossed and a card is drawn from a pack of 52 cards. Find the probability that the coin gives a head and the card is a King.

<u>Solution</u>. Let H = A Head is tossed, K = A King is drawn. Clearly H and K are independent. So
$$P(H \cap K) = P(H) P(K) \quad \text{(by independence)}$$
$$= \frac{1}{2} \cdot \frac{4}{52} = \frac{1}{26}.$$

28. <u>EXAMPLE</u>. (§20 revisited) A coin is tossed three times. Find the probability of three heads.

<u>Solution</u>. Let H_i = a head on the ith toss. (i = 1, 2, 3) We want $P(H_1 \cap H_2 \cap H_3)$. The tosses are independent.
So $P(H_1 \cap H_2 \cap H_3) = P(H_1) P(H_2 \cap H_3)$ (by independence)
$$= P(H_1) P(H_2) P(H_3) \quad \text{(by independence)}$$
$$= \left(\frac{1}{2}\right)^3 = \frac{1}{8}.$$

EXAMPLES TO DO: Page 33: Exs. 1 - 3.

29. **EXAMPLE.** Physical characteristics F and G occur in people of a certain race. Their respective probabilities for occurring in any particular person are 0.20 for F and 0.15 for G. They occur independently. Find the probability that a person chosen at random from the population

(a) has neither F nor G,
(b) has G given that he does not have F,
(c) has F given he has at least one of the two characteristics.

<u>Solution.</u> Let F and G denote the events that the person has F, G respectively. By independence,

$$P(F \cap G) = P(F) P(G) = (0.20).(0.15) = 0.03.$$

We now draw the Venn Diagram and mark in probabilities as shown, using $P(F \cap G)$ to start with, and subtracting it from the given values of $P(F)$ and $P(G)$ to find other entries. Then we proceed.

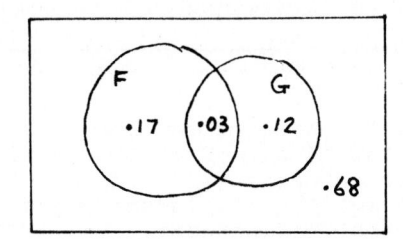

(a) P(neither F nor G) = 1 - 0.17 - 0.03 - 0.12 = 0.68.

(b) $P(G|\overline{F}) = \dfrac{P(G \cap \overline{F})}{P(\overline{F})}$ (using the formula in §18(ii))

$= \dfrac{0.12}{0.12 + 0.68} = \dfrac{12}{80} = \dfrac{3}{20}.$

(c) $P(F | F \cup G) = \dfrac{P(F \cap (F \cup G))}{P(F \cup G)}$ (using §18(ii) again)

$= \dfrac{P(F)}{P(F \cup G)} = \dfrac{0.20}{0.32} = \dfrac{5}{8}.$

<u>Notice</u>: (i) In doing questions like (b) and (c) here, the formula

$$P(B|A) = \dfrac{P(A \cap B)}{P(A)} \qquad \ldots (*)$$

from §18(ii) is extremely useful. You can alternatively think in part (c) for example of $P(F | F \cup G)$ as the fraction of the people in $(F \cup G)$ who have the characteristic F, i.e.

$$P(F | F \cup G) = \dfrac{0.17 + 0.03}{0.17 + 0.03 + 0.12} = \dfrac{5}{8}.$$

Be warned however that the formula (*) has much to commend it.

(ii) In (b) it is actually true that G and $\overline{F}$ are independent (see Ex. 4.9). So in fact

$$P(G|\overline{F}) = P(G) \quad \text{(from the result in §26)}.$$

This offers an alternative method.

EXAMPLES TO DO: Pages 33 - 34:   Exs. 4, 5, 9.

30. INDEPENDENCE OF THREE OR MORE EVENTS

Definition. Three events A, B, C are independent if and only if the following four conditions are met:

$$P(A \cap B) = P(A) P(B), \quad P(B \cap C) = P(B) P(C),$$

$$P(C \cap A) = P(C) P(A), \quad P(A \cap B \cap C) = P(A) P(B) P(C).$$

Notice: (i) None of these conditions can be deleted. In particular be warned that even if the three events are independent in pairs (i.e. the first three conditions are satisfied) then it is still possible that the three events are not independent.

(ii) Similar conditions involving all possible intersections of two or more events apply in the definition of independence of more than three events.

EXAMPLES TO DO: Page 34:   Exs. 7, 8.

31. CIRCUITS

Suppose that you are trying to send a message between two points A and B through a sequence of signalling stations. Successful transmission from A to B depends on successful transmission on each stage. To defend against loss or corruption of the message you could consider sending the same message by several different routes, in the hope that at least one route will succeed. This idea provides the background to the following type of example.

EXAMPLE. Suppose that two identical relay stations, which work independently and each of which has probability p of retransmitting a received signal forward correctly are connected between points X and Y. Find the probability that a signal passes correctly between X and Y when

(a) the relays are connected in series,
(b) the relays are connected in parallel.

Solution. (a) X —[1]—[2]— Y

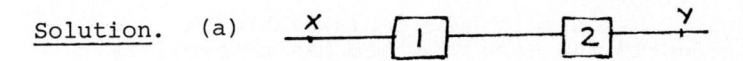

Let C = signal passes through relay 1.
Let D = signal passes through relay 2.

So P(signal on XY) = P(C ∩ D)
= P(C) P(D)   (by independence)
= p^2.

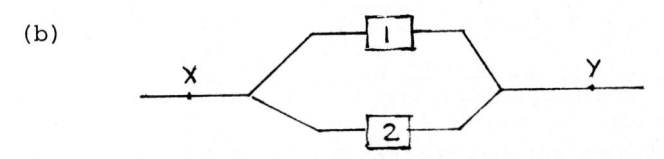

(b) [parallel circuit diagram with relays 1 and 2 between X and Y]

Here we find it easier (for reasons explained below) to calculate the probability of <u>no</u> signal between X and Y. We then subtract this value from 1 to give the required answer. So with the events C and D as above,

P(no signal on XY) = P((no signal on 1) ∩ (no signal on 2))

= P($\bar{C}$ ∩ $\bar{D}$)
= P($\bar{C}$) P($\bar{D}$)   (by independence)
= $(1-p)^2$.

So P(signal on XY) = $1 - (1-p)^2 = 2p - p^2$.

<u>Notice</u>: (i) In a <u>series situation</u> like (a), it is generally better to work out the probability of <u>successful transmission</u> directly. In a <u>parallel situation</u> it is generally better to work out the probability of <u>no transmission</u> (i.e. <u>failure</u>) and to subtract this value from 1. See §32 for another example of this.

In (b) it is possible here to adopt the direct approach and say

P(signal on XY) = P(C ∪ D) = P(C) + P(D) − P(C ∩ D)

= P(C) + P(D) − P(C) P(D) = $2p - p^2$.

This approach is reasonable here but rapidly becomes unmanageable as the number of components increases.

(ii) In (b) we used the fact that if C and D are independent then so also are $\bar{C}$ and $\bar{D}$. This is fairly clear intuitively, but Ex.4.9 indicates how to prove it.

(iii) §32 gives a harder example of this type.

32. EXAMPLE. Seven relay stations which work independently and each of which has probability p of forwarding a received signal lie between points X and Y on the electrical network shown:

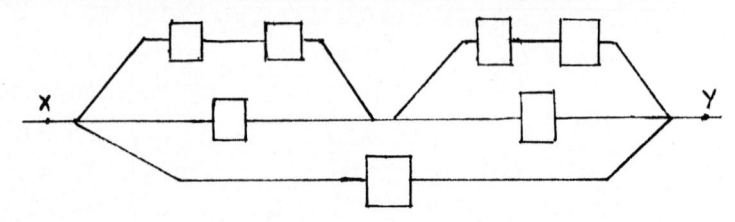

Find the probability that any given signal passes correctly from X to Y.

<u>Solution.</u>   The network is really

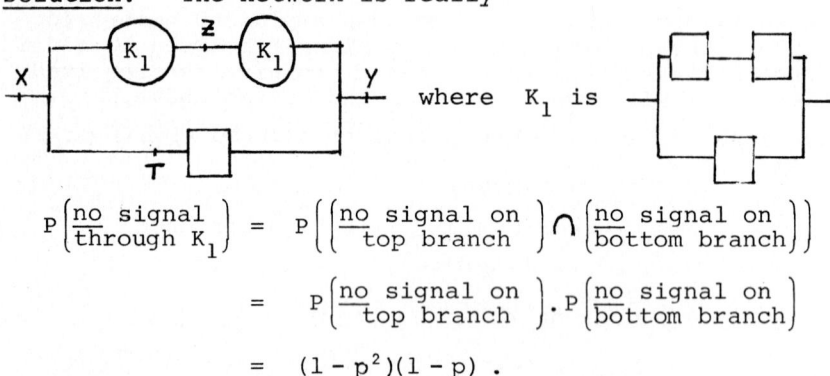

where K_1 is

$$P\left(\frac{\text{no signal}}{\text{through } K_1}\right) = P\left(\left(\frac{\text{no signal on}}{\text{top branch}}\right) \cap \left(\frac{\text{no signal on}}{\text{bottom branch}}\right)\right)$$

$$= P\left(\frac{\text{no signal on}}{\text{top branch}}\right) \cdot P\left(\frac{\text{no signal on}}{\text{bottom branch}}\right)$$

$$= (1-p^2)(1-p).$$

So   P(signal through K_1) = $1 - (1-p^2)(1-p) = q$, say.

So   P(signal on XZY) = q^2.

So   P(<u>no</u> signal on XZY) = $1 - q^2$.

Also   P(<u>no</u> signal on XTY) = $1 - p$.

So   P(<u>no</u> signal on XY) = P((<u>none</u> on XZY) $\cap$ (<u>none</u> on XTY))

$$= P(\underline{\text{none}} \text{ on } XZY) \cdot P(\underline{\text{none}} \text{ on } XTY)$$

$$= (1-q^2)(1-p).$$

So   P(signal on XY) = $1 - (1-q^2)(1-p)$,

where $q = 1 - (1-p^2)(1-p)$.

<u>Notice</u>: Here we have broken the circuit into components, calculating P(signal) for series situations and P(no signal) for parallel ones.

<u>EXAMPLES TO DO:</u>   Pages 33 - 34:   Ex. 6.

EXAMPLES 4

1. Events A, B, C are shown in the Venn diagram and the probabilities of lying in the disjoint pieces are marked in. State whether the following pairs of events are independent:

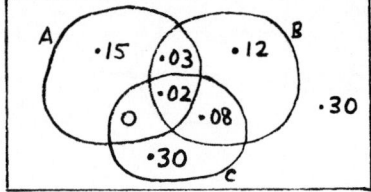

(i)   A   and   B,
(ii)   B   and   C,
(iii)   C   and   A,
(iv) $\bar{A}$ and $\bar{B}$,
(v) $\bar{A}$ and   B.

2. For the events A, B, C in Ex. 1 find
(i) $P(A \cup C \mid B)$,   (ii) $P(A \cap C \mid B)$,   (iii) $P(A \mid \bar{C})$,
(iv) $P(C \mid \bar{A})$,   (v) $P(C \mid A \cup B)$,   (vi) $P(C \mid A \cap B)$,
(vii) $P(\overline{A \cup B \cup C} \mid \overline{A \cup B})$,   (viii) $P(A \cup B \mid \overline{A \cap C})$.

3. A man and a woman each have a pack of cards. Each draws a card from his/her pack. Find the probability that

(a) the man draws a heart <u>and</u> the woman draws a club,
(b) the man draws a heart <u>or</u> the woman draws a club,
(c) both draw the same suit.

4. Events A and B are independent and $P(A) = 0.60$ and $P(A \cap B) = 0.12$. Find $P(B)$ and $P(A \mid \overline{A \cap B})$.

5. Two types of defect A and B occur in articles of a certain type. Defect A occurs with probability 0.4 while defect B occurs with probability 0.2. The defects occur independently. Find the probability that

(a) an article chosen at random has only one of the defects,
(b) an article has only one type of defect given it is defective,
(c) an article has no defect given it does not have defect B.

6. Relay stations, which work independently and each of which has probability p of forwarding a received signal correctly, lie in the various network arrangements shown. In each case find the probability of a signal being transmitted correctly from X to Y.

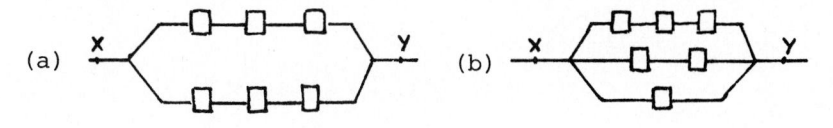

(a)   (b)

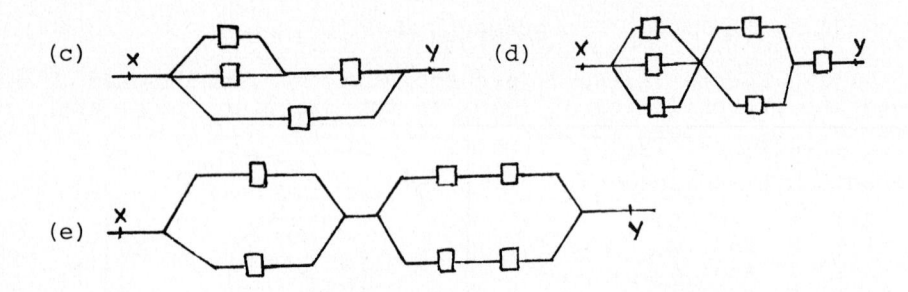

(c) [circuit diagram from x to y]   (d) [circuit diagram from x to y]

(e) [circuit diagram from x to y]

7. Let A, B be events which are <u>not</u> independent and let C = ∅ (the empty set). Use these events to illustrate that

$$P(X \cap Y \cap Z) = P(X)P(Y)P(Z)$$

is inadequate as a definition of the independence of three events X, Y, Z.

8. Events A, B, C are shown in the Venn diagram and the probabilities of lying in the disjoint pieces are marked in. Use these events to show that the first three conditions in the definition of independence of 3 events in §30 do not ensure the fourth condition. Explain in words why (in any reasonable sense) the 3 events A, B, C could not be regarded as independent.

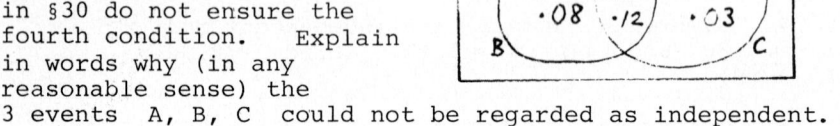

[Venn diagram with regions labeled: A only .15, outside .27, A∩B .20, A∩C .15, center 0, B only .08, B∩C .12, C only .03]

9. Prove that if C and D are independent events then so too are the pairs $\bar{C}$ and $\bar{D}$, and C and $\bar{D}$.

V. ONE DIMENSIONAL RANDOM VARIABLES IN GENERAL

33. Though the outcome of an experiment need not be a number (e.g. HHH in Experiment 2 in §2) we often want <u>numbers</u> as outcomes. A random variable X is <u>effectively a function</u> X from the sample space to the set of real numbers. In effect this means that, for each repetition of the experiment, X takes on a <u>real number value</u> depending on the outcome of the experiment.

<u>Example 1</u>. A coin is tossed four times. The random variable X is given by the number of heads that are obtained.

<u>Example 2</u>. A light bulb is burned until it burns out. The random variable Y is given by its life length in hours.

<u>Notice</u>: In Example 1, X can take the values 0, 1, 2, 3 and 4. In Example 2, Y can take all values in the set $\{a : a$ is a real number, $a \geq 0\}$, i.e. all values in the interval $[0, \infty)$.

34. DISCRETE AND CONTINUOUS

There are two basic types of random variable - <u>discrete</u> and <u>continuous</u>. The division between types depends on the <u>set of values</u> the random variable can take.

<u>Definition</u>. The random variable X is <u>discrete</u> if the number of values X can take is finite or countably infinite. (We say X has a <u>discrete distribution</u>.)

<u>Definition</u>. The random variable Y is <u>continuous</u> if the number of values Y can take is uncountably infinite. (We say Y has a <u>continuous distribution</u>.)

<u>Notice</u>: (i) In §33, X in Example 1 is discrete, while Y in Example 2 is continuous.

(ii) A <u>countably infinite</u> set is an infinite set in which the members can be written down in a sequence. For example, the set of all positive integers, namely, 1, 2, 3, ... is countably infinite. This contrasts with uncountably infinite sets like the closed interval [0, 1], i.e. the set of all real numbers between 0 and 1. Such uncountably infinite sets contain in some sense much "bigger" infinities of points than the infinities of points in countably infinite sets. An uncountable infinity cannot be written down in a sequence.

(iii) All the continuous random variables in this

book take all values in some subinterval of the real numbers, e.g. [0, 1], (5, 12) or even the set R of all real numbers itself. In the light bulb example in §33, the interval is [0, ∞).

(iv) There are some delicate questions in this area but the above definitions will serve well for the purposes of this book.

35. Most of the rest of this book deals with random variables. Particular types are the Binomial (Chapter 6) and the Poisson (Chapter 9), both of which are discrete, and the Normal (Chapter 10), which is continuous.

Chapter 7 looks at continuous random variables in detail, while Chapter 11 deals briefly with two dimensional random variables. Chapters 8 and 12 look at some ideas which apply to random variables in general.

VI. THE BINOMIAL DISTRIBUTION

36. THE BINOMIAL DISTRIBUTION

This is a standard discrete distribution.

<u>Definition</u>. Let E be an experiment and let A be an event which occurs with probability p - this probability being constant from one repetition of the experiment to the next. Suppose we perform n independent repetitions of E. Define a random variable X as the number of times that event A occurs in the course of the n repetitions. Then X is called a <u>binomial random variable with parameters n and p</u>. We also say that X has a <u>binomial distribution with parameters n and p</u>.

<u>Notice</u>: (i) If we toss a coin 6 times and let X denote the total number of heads obtained, then X is a binomial random variable with $n = 6$ and $p = \frac{1}{2}$.

(ii) Binomial random variables have the following two basic features:

(a) The outcome of the experiment can be regarded as "<u>success</u>" or "<u>failure</u>".
(b) The probability of "success" is constant (i.e. the same) on every repetition of the experiment.

The event A in the definition corresponds to "success" and the probability p corresponds to the constant probability of success on each repetition.

(iii) §37 gives the formula for the probability of k successes in a binomial experiment with parameters n and p.

37. THE FORMULA FOR THE BINOMIAL DISTRIBUTION

<u>RESULT</u>. Let X be a binomial random variable with parameters n and p, n being the number of repetitions and p being the constant probability of success. Then, for every integer k with $0 \leq k \leq n$,

$$P(X = k) = \binom{n}{k} p^k (1-p)^{n-k} .$$

<u>Proof</u>. We are interested in outcomes in which k successes (S) and (n-k) failures (F) occur in the n attempts. Think first of the case in which the outcome is first k successes and then (n-k) failures,

i.e. $\underbrace{S\,S\,S\ldots S}_{k}\underbrace{F\,F\,F\ldots F}_{(n-k)}$.

The probability of success on each repetition is p: so the probability of this particular sequence of successes and failures is $p^k(1-p)^{n-k}$.

Clearly however any other outcome in which there are k successes and $(n-k)$ failures also has the same probability $p^k(1-p)^{n-k}$. Furthermore the number of such outcomes is $\binom{n}{k}$, because in effect we must choose the k positions out of n in which the successes are to occur. So the total probability of k successes and $(n-k)$ failures is given by

$$\begin{pmatrix}\text{number of}\\ \text{cases}\end{pmatrix} \times \begin{pmatrix}\text{probability}\\ \text{of each case}\end{pmatrix},$$

i.e. $\binom{n}{k} p^k(1-p)^{n-k}$.

<u>Notice</u>: When $k = n$, the formula just reduces to p^n; similarly when $k = 0$, the formula just reduces to $(1-p)^n$.

38. **EXAMPLE.** Items coming off a production line each have probability $\tfrac{1}{3}$ of being top class. Find the probability that among 5 articles inspected there will be

(a) exactly 2 top class articles,
(b) no top class article,
(c) at most 2 top class articles,
(d) at least 2 top class articles.

<u>Solution.</u> Let X = No. of top class articles among the five.
The random variable X is binomial with $n = 5$ and $p = \tfrac{1}{3}$.

(a) $P(X = 2) = \binom{5}{2}\left(\tfrac{1}{3}\right)^2\left(\tfrac{2}{3}\right)^3 = \tfrac{80}{243}$. (Put $k = 2$ in §37.)

(b) $P(X = 0) = \binom{5}{0}\left(\tfrac{1}{3}\right)^0\left(\tfrac{2}{3}\right)^5 = \tfrac{32}{243}$. (Put $k = 0$ in §37.)

(c) $P(X \leq 2) = P(X = 0) + P(X = 1) + P(X = 2)$

$= \tfrac{32}{243} + \binom{5}{1}\left(\tfrac{1}{3}\right)^1\left(\tfrac{2}{3}\right)^4 + \tfrac{80}{243} = \tfrac{192}{243}$.

(d) $P(X \geq 2) = P(X = 2) + P(X = 3) + P(X = 4) + P(X = 5)$

$= 1 - P(X = 0) - P(X = 1)$

$= 1 - \tfrac{32}{243} - \tfrac{80}{243} = \tfrac{131}{243}$.

Notice: (i) This is a direct application of the result in §37. Effectively the 1 gram of "probability sand" (see §6) is spread over the values 0, 1, 2, 3, 4, 5 as shown in the diagram.

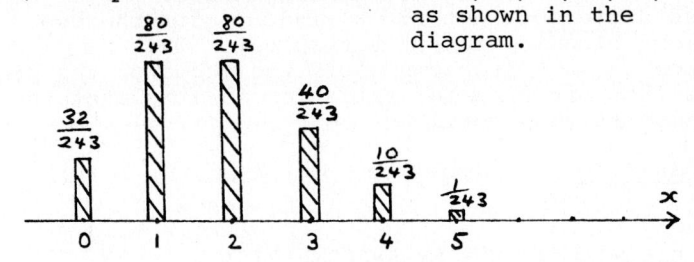

[Bar chart showing probabilities: $\frac{32}{243}$ at 0, $\frac{80}{243}$ at 1, $\frac{80}{243}$ at 2, $\frac{40}{243}$ at 3, $\frac{10}{243}$ at 4, $\frac{1}{243}$ at 5]

(ii) The technique in part (d) of finding $1 - P(\text{complementary event})$, which we have seen already elsewhere (e.g. §14), can shorten the work in many examples.

EXAMPLES TO DO: Pages 41 - 42: Exs. 1 - 4, 6.

39. EXAMPLE. A certain type of plane has 4 engines which work independently and each of which has probability z of failing on a certain flight. The plane can continue to fly provided that at least 3 engines are working; otherwise it has to make an emergency landing. Find the probability that

(a) on a given flight there is an emergency landing,
(b) in 10 flights there are at least 3 emergency landings.

Solution. Let X be the number of engines failing on the plane.
Then X is binomial with $n = 4$ and $p = z$.

(a) $P(\text{emergency landing}) = P(X \geq 2)$

$$= 1 - P(X=0) - P(X=1)$$
$$= 1 - (1-z)^4 - \binom{4}{1} z(1-z)^3$$
$$= 1 - (1-z)^4 - 4z(1-z)^3$$
$$= 1 - (1-z)^3(1-5z) = q \text{ say.}$$

(b) Let $Y = $ No. of emergency landings in 10 flights.
Then Y is binomial with $n = 10$, $p = q$.
So $P(\text{at least 3 emergency landings}) = P(Y \geq 3)$

$$= 1 - P(Y=0) - P(Y=1) - P(Y=2)$$
$$= 1 - (1-q)^{10} - \binom{10}{1}(1-q)^9 q - \binom{10}{2}(1-q)^8 q^2$$
$$= 1 - (1-q)^8(1 + 8q + 36q^2).$$

Notice: In part (b) of the above problem, the focus of the problem shifts from the number of engines failing to the number of emergency landings made. This is why we used <u>different</u> letters X and Y for the two different binomial random variables. This is good practice in such problems where the focus of the problem shifts from one binomial situation to a related one. Ignore it at your peril.

EXAMPLES TO DO: Pages 42 - 43: Exs. 5, 7, 10.

40. BINOMIAL VERSUS HYPERGEOMETRIC: PRODUCTION LINE VERSUS DWINDLING POPULATION

The following example and the comments below try to clarify the difference between these two types of situation. Inability to distinguish is a common source of trouble.

EXAMPLE. From an ordinary pack of 52 cards, 3 cards are drawn. Find the probability that these cards are 2 hearts and 1 non-heart, when the cards are drawn

(a) with replacement,
(b) without replacement.

Solution. Let X denote the number of hearts drawn.

(a) (<u>With replacement</u>)

The probability of drawing a heart is constant with the value 1/4. So, in this case, X is <u>binomial</u> with $n = 3$, $p = 1/4$ and $k = 2$ (in the formula of §37).

So $P(X = 2) = \binom{3}{2}(1/4)^2(3/4)^1 = \frac{9}{64} = 0.1406$.

(b) (<u>Without replacement</u>)

This is the <u>hypergeometric</u> situation described in §14.

(It is <u>not</u> binomial because the probability of drawing a heart varies from the drawing of one card to the next, depending on which cards have already been drawn. For the binomial, the probability of drawing a heart would have to be the same for each card drawn.)

The hypergeometric formula from §14 then gives

$$P(X = 2) = P(2H, 1 \text{ non-H}) = \frac{\binom{13}{2} \cdot \binom{39}{1}}{\binom{52}{3}} = \frac{117}{850} = 0.1376.$$

Notice: In (a) the probability of drawing a heart is constant at ¼ for every draw. This is the so called <u>production line situation</u> (illustrated in §38) where

the experiment can be repeated <u>indefinitely</u> but the
<u>probability of "success" remains constant</u>. The
binomial distribution then applies. Tossing a coin is
another example of a production line situation.

In (b) on the other hand, the probability of
drawing a heart varies as cards are drawn. This is the
<u>dwindling population situation</u>. At the start the
population consists of 52 cards but the number of cards
left dwindles as the experiment proceeds. Clearly such
an experiment <u>cannot</u> be repeated indefinitely because
after 52 draws the population has been exhausted. On
the first draw, the probability of drawing a heart is
certainly ¼ but if a heart is drawn on the first draw
then the probability of drawing a heart on the second
draw is down to 12/51, i.e. 0.235, because the number
of hearts left is 12 and the population is down to 51.
So the probability of drawing a heart is <u>not constant</u>
from one draw to the next. This dwindling population
situation is the classic setting for the <u>hypergeometric
distribution</u> and you should use the formula and
methods given in §14 and §15. The formula for the
binomial distribution <u>cannot</u> be used in this situation.

In doing problems be clear at the outset whether
you have a production line situation (i.e. p is constant
and the possibility of indefinite repetition) or
a dwindling population situation (i.e. p varies from one
attempt to the next and repetition only possible while
the population lasts out). So to summarise, we have:

PRODUCTION LINE = BINOMIAL

DWINDLING POPULATION = HYPERGEOMETRIC

<u>EXAMPLES TO DO</u>: Pages 42 - 43: Exs. 9, 8, 11 - 13.

EXAMPLES 6

1. A dice is tossed 4 times. Find the probability
of (a) exactly 2 sixes, (b) at least 2 sixes.

2. An unbiased coin is tossed 6 times. Find
the probability of
    (a) 2 heads and 4 tails, (b) no head,
    (c) an even number of heads.

3. Of the cars produced in a factory 20% are
defective. Each car is tested before being sold.
Eight cars are tested. Find the probability that
    (a) exactly six are defective,
    (b) at least six are defective,
    (c) at most six are defective.

4. Eggs are classified as either large or standard. The probability that an egg chosen at random will be large is 2/3. If six eggs are chosen at random, find the probability that

(a) exactly four are large,
(b) exactly four are large given that at least four are large.

If eggs are inspected one at a time, find the probability that the eighth egg inspected is the sixth large one found.

5. Boxes of cups contain four cups each. The probability that a random cup is defective is 1/3. Find the probability that

(a) in a random box there are exactly two defective cups,
(b) in a random box there are at least two defective cups,

Find the probability that among three randomly chosen boxes there is exactly one box with no defective cup.

6. The probability that an item coming off a production line is defective is p. Items are assembled in boxes of eight. A box is chosen at random. Find the probability that

(a) the box contains exactly two defective items,
(b) the box contains at least two defective items.

Denote your answer to (a) by q. If boxes are inspected one after another, find (in terms of q) the probability that the sixth box inspected is the fourth to contain exactly two defective articles.

7. Each electric motor coming off a production line has probability p of being defective. If five motors are inspected one after another, what is the probability of

(a) exactly two defective motors,
(b) at least three defective motors?

Motors are fitted into power units with four motors in each unit. Find the probability that if three units are inspected there is at least one unit with no defective motor.

8. An admirer is trying to see an heiress at a theatre. The theatre does six plays in the season, each play running for fifteen nights. The heiress goes to each play once, choosing her night at random. Find the probability that the admirer will be at the theatre on at least one night when the heiress is there if he goes to each play (a) once, (b) twice, (c) three times. Find the same probability if (d) he goes to the first six performances of the first play and forgets the rest, (e) if he goes to the first twelve performances

of the first play and forgets the rest.

9. An orchestra consists of 40 musicians of whom 20 are string players. They plan a series of 12 concerts and from past experience it is known that the probability that any given concert will be sold out is 1/3. Find the probability that at least 2 of the 12 concerts will be sold out.

On a particular day 5 members of the orchestra are hit by an illness which strikes at random. Find the probability the 5 struck are all string players.

10. In Ex.2.8 on page 16, find the probability that in a period of seven nights there will be at least one night on which there is no locomotive with snowplough in the town.

11. A company has three factories A, B, C making 20%, 70% and 10% of its output respectively. The percentages of articles made by A, B, C that are defective are 10%, 20%, 30% respectively. An article is chosen at random from stock and is found to be defective. Find the probability that it came from factory C.

Each factory packs its articles into identical boxes and the boxes are taken to the company store. An inspector there chooses a box at random, inspects three articles from it and finds exactly one to be defective. Find the probability that this box came from factory A.

12. Three shops X, Y, Z sell 30%, 50%, 20% of the shirts in a certain town. On average it is found that among the shirts sold by the shops X, Y, Z, 40%, 10%, 5% are substandard, respectively.

A man buys one shirt in each shop. Find the probability he receives exactly one substandard shirt.

A woman buys two shirts in shop X and two in shop Y. Find the probability she receives exactly two substandard shirts.

13. A company has three factories A, B, C all making chairs, the outputs of A, B, C being 10%, 50% and 40% of the total output, respectively. The percentages of the outputs of A, B, C that are defective are 10%, 20% and 5% respectively. A set of three chairs is assembled. Find the probability that the set contains exactly one defective chair if
  (a) all the chairs come from factory A,
  (b) there is one chair from each factory.
At the company store the chairs from all the factories are pooled. Chairs in stock at the store are then inspected one by one. Find the probability that the fifth chair inspected is the second defective found.

VII. CONTINUOUS RANDOM VARIABLES

41. PROBABILITY DENSITY FUNCTION FOR A CONTINUOUS RANDOM VARIABLE

The two basic types of random variable - discrete and continuous - were defined in §34. For a <u>discrete</u> random variable, we can often draw a probability distribution in histogram form (as in §38), e.g.

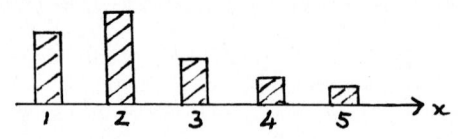

where the heights of the columns represent the amount of "probability sand" (from the 1 gram total) (see §6) associated with that value of X.

Our prototype <u>continuous</u> random variable is the life length X hours of a light bulb. In this case we have, instead of a histogram, a <u>probability density function</u> f, where the area under the curve $y = f(x)$ in

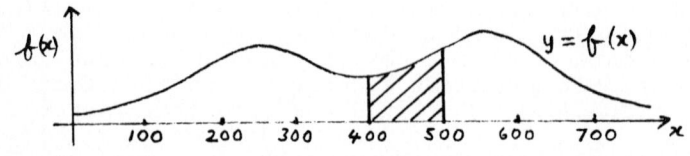

the range $400 \leq x \leq 500$ represents the amount of "probability sand" associated with light bulbs with life length between 400 and 500 hours.

Notice: (i) Recall from §34 that for a continuous <u>random</u> variable X we can think for our purposes of the values taken by X as making up an interval in the real numbers. For the light bulb experiment the interval is $[0, \infty)$. This means that we cannot use the histogram approach because the values of X are so tightly packed together.

(ii) The total area under the graph of the probability density function (i.e. between the function and the x-axis) is 1 unit. This corresponds to the total mass of "probability sand" being 1 gram.

(iii) For a function f to be the <u>probability density function</u> (p.d.f.) of a continuous random variable X, we require the following three conditions:

(a) $f(x) \geq 0$ for all real numbers x;

(b) $\int_{-\infty}^{\infty} f(x) \, dx = 1$;

(c) For every pair of real numbers c, d with c ≤ d,
$$P(c \le X \le d) = \int_c^d f(x)\,dx.$$

In practice the range $-\infty$ to ∞ in (b) will degenerate to the interval on which X takes values. For example, if X takes only values on [3, 7] then condition (b) becomes
$$\int_3^7 f(x)\,dx = 1.$$

In such cases we regard f as taking the value 0 on $(-\infty, 3) \cup (7, \infty)$. This means that (b) still applies though the contributions to the integral from the ranges $(-\infty, 3)$ and $(7, \infty)$ are zero.

42. EXAMPLE. Batteries have life length X hours, where X is a continuous random variable taking values in the range $100 \le X \le 200$ and with the p.d.f.

$$f(x) = \begin{cases} \dfrac{a}{x^2} & (100 \le x \le 200), \\ 0 & \text{otherwise.} \end{cases}$$

Find (i) the constant a,
  (ii) $P(X \ge 150)$, i.e. the probability that a battery lasts at least 150 hours,
  (iii) $P(X \ge 175 \mid X \ge 150)$, i.e. the probability that a battery lasts at least 175 hours given it lasts at least 150 hours.

Also, if three batteries are tested independently, find the probability that at least one is still working after 150 hours.

Solution. (i) The constant a is just a scale factor which we find by using the fact that the total probability available is 1. So
$$\int_{100}^{200} \frac{a}{x^2}\,dx = 1,$$
i.e. $a\left[\dfrac{-1}{x}\right]_{100}^{200} = 1,$

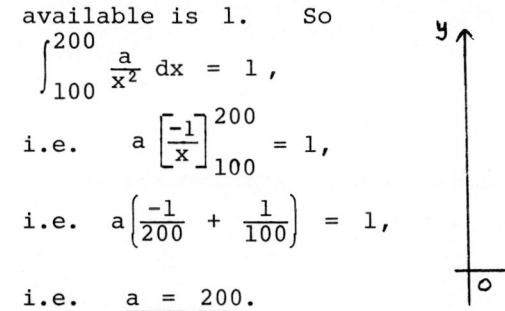

i.e. $a\left(\dfrac{-1}{200} + \dfrac{1}{100}\right) = 1,$

i.e. $a = 200.$

(ii) Setting $a = 200$ in the p.d.f. we now proceed.
$$P(X \ge 150) = P(150 \le X \le 200) = \int_{150}^{200} \frac{200}{x^2}\,dx = \left[\frac{-200}{x}\right]_{150}^{200} = \frac{1}{3}.$$

(iii) $P(X \geq 175 \mid X \geq 150) = \dfrac{P((X \geq 175) \cap (X \geq 150))}{P(X \geq 150)}$

(by §18(ii))

$= \dfrac{P(X \geq 175)}{P(X \geq 150)} = \dfrac{1}{1/3} \int_{175}^{200} \dfrac{200}{x^2} \, dx = \dfrac{3}{7}$.

(Last part) This is binomial. Let Y denote the number of batteries still working after 150 hours. Here n = 3, p = ⅓ and we want $P(Y \geq 1)$. So

$P(Y \geq 1) = 1 - P(Y = 0)$

$= 1 - \binom{3}{0}(1/3)^0 (2/3)^3 = 1 - \dfrac{8}{27} = \dfrac{19}{27}$.

Remarks: (i) The constant a is determined using the fact that the total area under the graph of the p.d.f. is 1. In a real life situation you might know the form of the p.d.f. (i.e. you might know that it tailed off as $1/x^2$); you would then determine the scale factor a in the above manner.

(ii) The formula

$$P(B \mid A) = \dfrac{P(A \cap B)}{P(A)} \qquad \ldots (*)$$

from §18(ii) is very useful in questions like (iii). (Compare also §29 and §45.) Alternatively you could think of it as

$P(X \geq 175 \mid X \geq 150) = \dfrac{\text{mass of "sand" beyond 175}}{\text{mass of "sand" beyond 150}}$.

Such thinking gives good insight, but the method using the conditional probability formula (*) is clinically safe in all such situations.

(iii) The amount of probability associated with one particular point <u>for a continuous distribution</u> is zero: in the above example,

$P(X = 150) = \int_{150}^{150} \dfrac{200}{x^2} \, dx = 0$.

Similarly notice that <u>for a continuous distribution</u>

$P(c \leq X \leq d) = P(c < X < d)$.

So, in the above example,

$P(X > 150) = P(X \geq 150) = \dfrac{1}{3}$.

EXAMPLES TO DO: Pages 50 - 51: Exs. 1, 2, 3.

43. CUMULATIVE DISTRIBUTION FUNCTION

This idea applies both to discrete and continuous distributions. Suppose first that we have a discrete distribution, e.g. the binomial with $n = 5$ and $p = \frac{1}{2}$ for which the distribution of probability is shown:

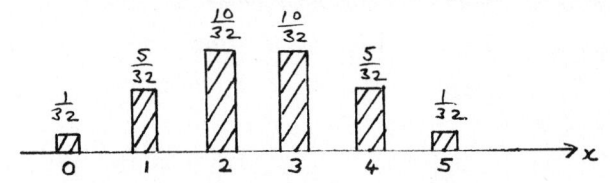

[Bar chart showing probabilities $\frac{1}{32}, \frac{5}{32}, \frac{10}{32}, \frac{10}{32}, \frac{5}{32}, \frac{1}{32}$ at $x = 0, 1, 2, 3, 4, 5$]

It is often useful to have a running total of the probability which accumulates as we go from $X = 0$ to $X = 5$. So, in this case, we find

$P(X \leq -1) = 0$; $\quad P(X \leq 0) = \frac{1}{32}$; $\quad P(X \leq \frac{1}{2}) = \frac{1}{32}$;

$P(X \leq 1) = \frac{6}{32}$; $\quad P(X \leq 2) = \frac{16}{32}$; $\quad P(X \leq 2\frac{1}{2}) = \frac{16}{32}$;

$P(X \leq 4\frac{1}{2}) = \frac{31}{32}$; $\quad P(X \leq 5) = 1$; $\quad P(X \leq 8) = 1$.

The function which records this running total is the cumulative distribution function (c.d.f.).

Definition. Let X be a random variable, discrete or continuous. Then the cumulative distribution function of X is the function $F: \mathbb{R} \to [0, 1]$ defined by

$$F(a) = P(X \leq a) \quad \text{for every } a \in \mathbb{R}.$$

Notice: (i) Since $F(a)$ is the running total of the mass of "probability sand" to the left of the point a, it is fairly clear that F is an increasing function, increasing from the value 0 up to the value 1. (Look at the running total in the above example.)

(ii) For a discrete random variable the c.d.f. is found by summing up the probabilities as in the above example.

(iii) For a continuous random variable, the c.d.f. is the integral of the p.d.f.. To see this remember that

$$F(a) = P(X \leq a) = \int_{-\infty}^{a} f(x)\, dx .$$

So, for a continuous random variable, to find the c.d.f. from the p.d.f. you integrate, and to find the p.d.f. from the c.d.f. you differentiate.

44. **EXAMPLE.** (Continuation of the example in §42)
Let X be the random variable given in §42 on the life length of batteries. Find the c.d.f. of X.

Solution. Denote the c.d.f. by F.
Then, for a with $100 \leq a \leq 200$,

$$F(a) = \int_{100}^{a} \frac{200}{x^2} dx = \left[-\frac{200}{x} \right]_{100}^{a} = 2 - \frac{200}{a}.$$

So the c.d.f. is given by

$$F(x) = \begin{cases} 0 & (x < 100), \\ 2 - \frac{200}{x} & (100 \leq x \leq 200), \\ 1 & (x > 200). \end{cases}$$

Notice: The c.d.f. F increases from the value 0 at $x = 100$ to the value 1 at $x = 200$. This corresponds to the running total of "probability so far" increasing from 0 up to the maximum possible value of 1. It is useful to include the values 0 for $x < 100$ and 1 for $x > 200$ for later results.

EXAMPLES TO DO: Page 51: Ex. 4.

45. **THE EXPONENTIAL DISTRIBUTION**

The following example illustrates this standard distribution.

EXAMPLE. The continuous random variable X takes all values in the interval $[0, \infty)$ and has p.d.f. given by

$$f(x) = \begin{cases} a e^{-ax} & (x \geq 0), \\ 0 & (\text{otherwise}), \end{cases}$$

where a is a positive constant. Check that this is indeed a valid p.d.f.. Determine
(i) $P(X < 1)$, (ii) $P(X < 3 \mid X < 1)$, (iii) the c.d.f. of X.

Solution. To see that f is a valid p.d.f., notice first that $f(x) \geq 0$ for all values of x. Also

$$\int_0^\infty a e^{-ax} dx = [-e^{-ax}]_0^\infty = 1,$$

as is required for a p.d.f.. So we conclude that f is a valid p.d.f., no matter what the value of a (see the remark below). We then take the other parts as follows.

(i) $P(X < 1) = \int_0^1 a e^{-ax} dx = [-e^{-ax}]_0^1 = 1 - e^{-a}$.

(ii) $P(X < 3 \mid X > 1) = \dfrac{P((X < 3) \cap (X > 1))}{P(X > 1)}$ (by §18(ii))

$= \dfrac{P(1 < X < 3)}{P(X > 1)}$

$= \dfrac{e^{-a} - e^{-3a}}{e^{-a}} = 1 - e^{-2a}$.

(iii) Denote the c.d.f. by F. Then, for $b \geq 0$,

$F(b) = P(X \leq b) = \int_0^b a e^{-ax} dx = [-e^{-ax}]_0^b = 1 - e^{-ab}$.

So the c.d.f. is given by

$$F(x) = \begin{cases} 0 & (x < 0), \\ 1 - e^{-ax} & (x \geq 0). \end{cases}$$

Remarks. (i) The constant a in the p.d.f. in this example is not to be found. (Contrast §42.) Here the p.d.f. integrates up to 1 for every value of a. Different values of a correspond to different rates of exponential decline: one real life situation might be described by having a = 100, while another might be described by having a = 4. This point arises in further examples later.

(ii) Notice the use of the letter b in part (iii), rather than a in the definition of c.d.f. in §43, in order to avoid confusion with the a already in this question. If necessary defend your work against such confusion by choosing your notation carefully. A similar point was mentioned with X and Y in §39.

46. THE UNIFORM DISTRIBUTION

This is another standard distribution which we illustrate with an example.

EXAMPLE. The length (in inches) of offcuts of wood in a sawmill can be regarded as a continuous random variable X, with the p.d.f. given by

$$f(x) = \begin{cases} \dfrac{1}{12} & (0 \leq x < 12), \\ 0 & \text{otherwise}. \end{cases}$$

Find (a) $P(3 < X < 6)$, (b) the c.d.f. of X.

Solution. (a) $P(3 < X < 6) = \int_3^6 \frac{1}{12} dx = [x/12]_3^6 = \frac{1}{4}$.

(b) For $0 \leq a \leq 12$, $P(X \leq a) = \int_0^a \frac{1}{12} dx = \frac{a}{12}$.

So the c.d.f. is F given by

$$F(x) = \begin{cases} 0 & (x < 0), \\ \frac{x}{12} & (0 \leq x \leq 12), \\ 1 & (x > 12). \end{cases}$$

Notice: This illustrates a distribution for which the p.d.f. is flat on the interval [0, 12]. The height of the distribution is 1/12 to make the total area under the graph of the p.d.f. equal to 1. In general the uniform distribution on interval [a, b] has the p.d.f.

$$f(x) = \begin{cases} \frac{1}{(b-a)} & (a \leq x \leq b), \\ 0 & \text{otherwise.} \end{cases}$$

This distribution occurs naturally in many situations. Try to think of some yourself.

EXAMPLES 7

1. The probability density function of a random variable X is given by

$$f(x) = \begin{cases} ax^2(3-x) & (0 \leq x \leq 3), \\ 0 & \text{otherwise.} \end{cases}$$

Find (i) the constant a, (ii) $P(X \leq 1)$, (iii) $P(X \leq \frac{1}{2} \mid X \leq 1)$.

2. The probability density function of a continuous random variable X is given by

$$f(x) = \begin{cases} ax(3x+2) & (0 \leq x \leq 4), \\ 0 & \text{otherwise.} \end{cases}$$

Find (i) the constant a, (ii) $P(X \geq 1)$, (iii) $P(X < 3 \mid X > 2)$.

If four independent observations of X are made, what is the probability that exactly two of them are greater than 2?

3. The p.d.f. of a continuous random variable X is given by
$$f(x) = \begin{cases} a(16x - x^3) & (0 < x < 4), \\ 0 & \text{otherwise.} \end{cases}$$
Find (i) the constant a, (ii) $P(X > 1)$, (iii) $P(X > 2 \mid X > 1)$.

Observations of X are made until two successive observations are greater than 1. What is the probability that exactly five observations are made?

4. A continuous random variable X has the p.d.f. f given by
$$f(x) = \begin{cases} 0 & (x < 4), \\ 4/x^2 & (x \geq 4). \end{cases}$$
Find (i) the cumulative distribution of X, (ii) $P(X > 6)$, (iii) $P(X < 10 \mid X > 6)$.

VIII. EXPECTED VALUE AND VARIANCE OF A RANDOM VARIABLE

47. In a general talk you may wish to give an impression of the behaviour of some random variable without giving the full details of its probability distribution (if it is discrete) or its probability density function (if it is continuous). [You may yourself not know the probability density function.] Stating the expected value and the variance of the random variable - they are both just numbers - will provide your audience with such an impression.

The <u>expected value</u> (or <u>mean</u>) of the random variable is effectively its average value (in a sense defined below). The <u>variance</u> of the random variable is a non-negative number which gives an idea of how widely spread the observations of the random variable are likely to be - the larger the variance, the more scattered the observations.

EXPECTED VALUE

48. EXPECTED VALUE - DISCRETE CASE

The following example attempts to motivate the definition below.

<u>EXAMPLE 1.</u> A normal dice is thrown. X denotes the outcome of a throw. Find the average value of X.

<u>Solution.</u> The average value is given by

$$\frac{1 + 2 + 3 + 4 + 5 + 6}{6} = 3\tfrac{1}{2}.$$

<u>Notice:</u> The average value can also be written as

$$1 \cdot \tfrac{1}{6} + 2 \cdot \tfrac{1}{6} + 3 \cdot \tfrac{1}{6} + 4 \cdot \tfrac{1}{6} + 5 \cdot \tfrac{1}{6} + 6 \cdot \tfrac{1}{6},$$

which is more in line with the definition below.

<u>EXAMPLE 2.</u> A special dice has one face showing 6, three faces showing 5 and two faces showing 1. All faces are equally likely. X denotes the outcome of a throw. Find the average value of X.

<u>Solution.</u> The average value is given by

$$\frac{6+5+5+5+1+1}{6} = 6 \cdot \tfrac{1}{6} + 5 \cdot \tfrac{3}{6} + 1 \cdot \tfrac{2}{6} = \tfrac{23}{6}.$$

<u>Notice:</u> This points forward to the following definition.

Definition. Let X be a discrete random variable with possible values $x_1, x_2, x_3, \ldots$, and let $p(x_i)$ denote $P(X = x_i)$. Then the **expected value of** X (denoted by $E(X)$) is defined by

$$E(X) = \Sigma\, x_i\, p(x_i).$$

Notice: (i) The summation sign Σ in the definition tells you to sum all possible terms of the given form. Often the sum is a sum of finitely many terms. If the sum consists of infinitely many terms then caution about convergence may be needed, but this is beyond the scope of this book.

(ii) $E(X)$ can also be called the **mean value** of X.

(iii) In Example 1 above, $E(X) = 3\frac{1}{2}$, while in Example 2, $E(X) = 23/6$.

(iv) The definition effectively takes a weighted average value of X, taking account of how much probability is attached to each x-value.

49. EXAMPLE. A darts player aims to score 60 (i.e. a treble twenty) with each dart he throws. The diagrams show on the left the scores in the beds of the board surrounding the treble twenty, and on the right his probabilities of hitting these beds.

5	20	1
15	60	3
5	20	1

·06	·15	·06
·10	·25	·10
·06	·15	·06

There is also probability 0.01 of bouncing off and scoring zero. Find his expected score from one dart.

Solution. $E(\text{Score}) = (60 \times 0.25) + 2(20 \times 0.15)$
$\qquad\qquad + (15 \times 0.10) + (3 \times 0.10)$
$\qquad\qquad + 2(5 \times 0.06) + 2(1 \times 0.06)$
$\qquad\qquad + (0 \times 0.01)$
$\qquad = 15 + 6 + 1.5 + 0.3 + 0.6 + 0.12$
$\qquad = 23.52.$

EXAMPLES TO DO: Page 64: Exs. 1, 2.

50. EXPECTED VALUE - CONTINUOUS CASE

Corresponding to the definition in the discrete case in §48, we give the following definition for the continuous case.

<u>Definition</u>. Let X be a continuous random variable with probability density function f. Then the <u>expected value of X</u> (denoted by E(X)) is defined by

$$E(X) = \int_{-\infty}^{\infty} x f(x) \, dx,$$

provided that this integral converges.

<u>Notice</u>: (i) There is a clear analogy between the form of the definition in the discrete case in §48 and this definition for the continuous case.

(ii) Though the range of integration is given as $(-\infty, \infty)$ in practice it degenerates to the set of values for which $f(x)$ is non-zero, e.g. to [100, 200] in the example on batteries in §42 or to $[0, \infty)$ in the example in §51.

(iii) It is possible for the integral in the definition to diverge (e.g. Ex.8.17 at the end of this chapter). This will not concern us much in this book.

(iv) The following example gives credence to the above definition.

<u>EXAMPLE</u>. The length of a bus journey taken by passengers on a certain route is X km, where X is uniformly distributed on the interval [0, 20]. Find E(X).

<u>Solution</u>. Here $f(x) = \frac{1}{20}$ $(0 \leq x \leq 20)$. (See §46.)

Then from above, $E(X) = \int_0^{20} \frac{x}{20} \, dx = \left[\frac{x^2}{40}\right]_0^{20} = 10,$

as you would expect.

51. <u>EXAMPLE</u>. Electrical components have life length X years, where X is a continuous random variable with the p.d.f.

$$f(x) = \begin{cases} \frac{1}{2} e^{-\frac{1}{2}x} & (x > 0), \\ 0 & \text{otherwise.} \end{cases}$$

Find the expected life of a component.

<u>Solution</u>. $E(X) = \int_0^{\infty} x f(x) \, dx = \int_0^{\infty} \frac{1}{2} x e^{-\frac{1}{2}x} \, dx$

$$= \int_0^\infty 2ue^{-u} \, du \quad \text{(on setting } u = \tfrac{1}{2}x\text{)}$$

$$= 2\left[-(u+1)e^{-u}\right]_0^\infty \quad \text{(on integrating by parts)}$$

$$= 2[0 + 1] = 2.$$

So the expected life of a component is 2 years.

Notice: Here the integral is amenable to integration by parts. Be warned however that in some problems where the p.d.f. involves an exponential over $[0, \infty)$ or $(-\infty, \infty)$, integration by parts can be very long. In such cases the Gamma function (§100) may then shorten the work. In still other cases, integration by parts is impossible and the Gamma function or a substitute for it cannot then be avoided (e.g. in §61).

EXAMPLES TO DO: Pages 64 - 67: Exs. 3, 17 (for $E(X)$).

52. EXPECTED VALUE OF A FUNCTION OF A RANDOM VARIABLE

Think of a car dealer who sells X cars per week, where X is a random variable. Related to X is $G(X)$ his weekly profit. He may be interested in $E(X)$ but he is likely to be even more interested in $E(G(X))$.

The following result deals with such a situation, where X is a given random variable and we wish to calulate $E(Y)$ where $Y = \phi(X)$ is a function of X.

RESULT. Let X be a random variable and let $Y = \phi(X)$. (So Y is another random variable and Y is a function of X.) Then

(a) if X is discrete,

$$E(Y) = \Sigma \, \phi(x_i) p(x_i),$$

where $p(x_i)$ denotes $P(X = x_i)$;

and (b) if X is continuous,

$$E(Y) = \int_{-\infty}^\infty \phi(x) f(x) \, dx.$$

Notice: (i) Be clear that it is the probabilities of the various values of the original random variable X that appear in (a) and it is the p.d.f. f of the original random variable X that occurs in (b).

(ii) We do not prove this result but as some justification look at the following argument relating to part (a):

$$E(Y) = \Sigma\, y_k P(Y = y_k) \quad \text{(summing over all possible } Y \text{ values)}$$

$$= \Sigma\, y_k \cdot \begin{pmatrix} \text{sum of the probabilities of all} \\ x_i \text{ values with } \phi(x_i) = y_k \end{pmatrix}$$

$$= \Sigma\, \phi(x_i) p(x_i) \quad \text{(summing over all possible } X \text{ values)}.$$

(iii) §53 - 55 give examples of the use of this result.

53. <u>EXAMPLE</u>. The current X amps in a 6 ohm resistor is a uniformly distributed random variable with $1 \leq X \leq 3$. The power Z watts in the resistor is then given by $Z = 6X^2$. Find the expected value of Z.

<u>Solution</u>. The p.d.f. of X is given by $f(x) = \frac{1}{2}$ for $1 \leq x \leq 3$. So, by the result of §52,

$$E(Z) = E(6X^2) = \int_1^3 6x^2 \cdot \tfrac{1}{2}\, dx = \left[x^3\right]_1^3 = 26.$$

<u>Notice</u>: At first sight you might think that the answer here should be 24, on the basis that the average value of the current is 2 and $6 \cdot 2^2 = 24$. This is wrong. For some insight into why it is wrong, notice that the power Z can take values in the interval [6, 54] and that the probability of lying in each of the intervals [6, 24] and [24, 54] is $\frac{1}{2}$.

54. <u>EXAMPLE</u>. In a shooting gallery a man fires shots at a circular target of radius 3 units. The target is split into a central disc and two surrounding rings by circles of radii 1 and 2 units as shown. The scores are 10 points for a hit in the central disc, 6 points for a hit in the inner ring, 3 points for a hit in the outer ring and zero for a shot which misses the target. The distance from the centre of the target (X units) at which a shot strikes the plane of the target is a random variable with p.d.f. given by

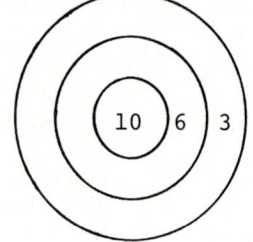

$$f(x) = \frac{1}{(1 + x)^2} \qquad (x \geq 0).$$

Find the expected score from one shot.

Solution. E(Score) = Σ(Score).(Probability of that score)

$$= 10.P(0 \leq X \leq 1) + 6.P(1 \leq X \leq 2) + 3.P(2 \leq X \leq 3)$$

$$= 10 \int_0^1 \frac{dx}{(1+x)^2} + 6 \int_1^2 \frac{dx}{(1+x)^2} + 3 \int_2^3 \frac{dx}{(1+x)^2}$$

$$= 10 \left[\frac{-1}{1+x}\right]_0^1 + 6 \left[\frac{-1}{1+x}\right]_1^2 + 3 \left[\frac{-1}{1+x}\right]_2^3$$

$$= 5 + 1 + \tfrac{1}{4} = 6\tfrac{1}{4}.$$

55. EXAMPLE. The life length of an electric motor is a continuous random variable X years with p.d.f. given by

$$f(x) = \begin{cases} \tfrac{1}{9}x(4-x) & (0 \leq x \leq 3), \\ 0 & \text{otherwise.} \end{cases}$$

(Note that this means no motor survives more than three years.) It costs the manufacturer £A to make each motor and he sells it for £B. He does however guarantee a partial refund of £C on motors with life length less than one year. Find the manufacturer's expected profit on a motor.

Solution. The manufacturer's profit (gain) on a motor is £G(X), where G is a function of the life length X. In fact,

$$G(x) = \begin{cases} B - A & (1 \leq x \leq 3), \\ B - A - C & (0 \leq x < 1). \end{cases}$$

So the expected profit is

$$E(G) = (B-A).P(1 \leq X \leq 3) + (B-A-C).P(0 \leq X < 1)$$

$$= (B-A) \int_1^3 \tfrac{1}{9}x(4-x)\,dx + (B-A-C) \int_0^1 \tfrac{1}{9}x(4-x)\,dx$$

$$= (B-A) \left[\tfrac{1}{9}(2x^2 - \tfrac{x^3}{3})\right]_1^3 + (B-A-C) \left[\tfrac{1}{9}(2x^2 - \tfrac{x^3}{3})\right]_0^1$$

$$= \tfrac{22}{27}(B-A) + \tfrac{5}{27}(B-A-C) = (B-A) - \tfrac{5}{27}C.$$

Notice: Other methods of reaching this answer are possible, but the method illustrated, in which the gain in clearly written down in the various cases at the start is well worth remembering.

EXAMPLES TO DO: Pages 65 - 66: Exs. 5 - 11.

56. HEREDITARY RESULTS FOR EXPECTED VALUE

The following results give the expected values of scalar multiples, sums and products of given random variables.

RESULT 1. Let X, Y be random variables and let k be a constant. Then

(i) $E(k) = k$;

(ii) $E(kX) = kE(X)$;

(iii) $E(X+Y) = E(X) + E(Y)$.

Also, <u>provided that X and Y are independent</u>,

(iv) $E(XY) = E(X)E(Y)$.

RESULT 2. Let $X_1, X_2, \ldots, X_n$ be random variables. Then $E(X_1 + X_2 + \ldots + X_n) = E(X_1) + E(X_2) + \ldots + E(X_n)$.

Notice: (i) These results are of course subject to the existence of the component expected values: we are not treating cases like those mentioned in §50(iii).

(ii) Result 2 is just an extension of Result 1(iii).

(iii) In Result 1(iv) it is imperative <u>that X and Y be independent</u>. The example of §53 in which

$$E(6X^2) \neq 6(E(X))^2$$

shows that this condition cannot be dropped.

VARIANCE AND STANDARD DEVIATION

57. THE IDEA OF VARIANCE

The expected value of a random variable X tells you the "average value" of X but it tells you nothing about how closely concentrated round the expected value the distribution of probability is. For example, for light bulb life length, consider the following distributions all with expected value 500.

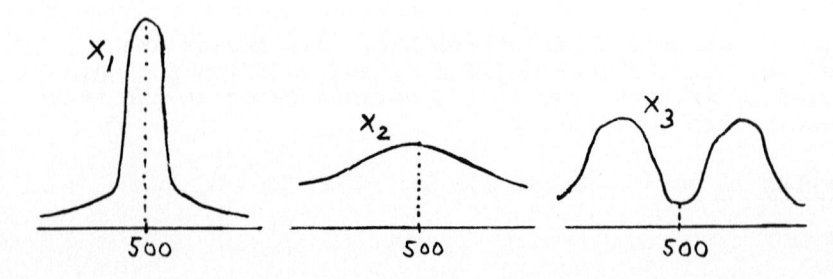

Here $E(X_i) = 500$ in each case but the actual distributions of probability are very different.

To give a measure of whether the probability distribution of X is closely concentrated <u>round the expected value</u> or widely spread <u>away from the expected value</u> we introduce a number called the <u>variance of</u> X (denoted by V(X)). In all cases $V(X) \geq 0$ and the larger V(X), the more widely spread the distribution. So in the three cases mentioned above, we would expect

$$V(X_1) < V(X_2) < V(X_3).$$

In a vague way you can think of the variance as the moment of inertia of the distribution round the mean.

58. VARIANCE AND STANDARD DEVIATION

Here are the definitions.

<u>Definition.</u> Let X be a random variable. The <u>variance of</u> X is defined by

$$V(X) = E((X - E(X))^2).$$

The <u>standard deviation of</u> X is defined to be $\sqrt{(V(X))}$. The <u>standard deviation</u> is usually denoted by the Greek letter σ (sigma). If necessary we can write σ_X to emphasise it relates to X.

<u>Notice</u>: (i) Recall that the square of a real number is always non-negative, i.e. for every real number a, we have $a^2 \geq 0$. So the values taken by the random variable $(X - E(X))^2$ are non-negative, and consequently the expected value of $(X - E(X))^2$ is non-negative, i.e. V(X) is non-negative, i.e. $V(X) \geq 0$.

(ii) The definition of variance effectively calculates the average value of the <u>square</u> of the differences $X - E(X)$, (i.e. the difference of an observation from the mean). The differences are squared to prevent positive differences cancelling negative ones in the averaging proccess.

The <u>standard deviation</u> (as the square root of the variance) attempts to undo the squaring process in some sense, and hence gives some sort of measure of deviation from the mean for the random variable.

(iii) Suppose we have a random variable, discrete or continuous, with mean μ (i.e. expected value μ) and standard deviation σ. There are several results which count off standard deviations on each side of the mean and then make statements about the probability of an observation lying within so many standard deviations of the mean.

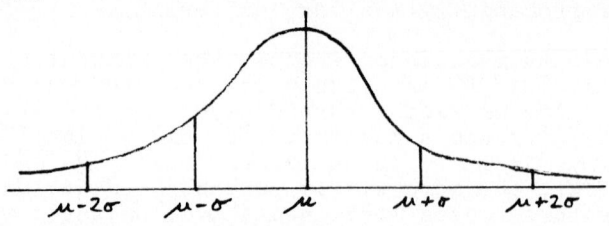

For example, for the normal distribution (see Remark(ii) in §81), notice that roughly   95% of the probability lies within two standard deviations of the mean. Chebyshev's Inequality (§92) gives another result of this general type.

59.   As explained in §58(ii), the form of the definition of the variance, namely

$$V(X) = E((X - E(X))^2), \qquad \ldots (*)$$

certainly has the look of achieving what it sets out to do - i.e. to measure how scattered from the mean the observations of X are likely to be.   However the formula (*) can be quite awkward to use for calculating the variance where the p.d.f. is given.   The following result gives an equivalent form of (*) which in many cases is much easier to use.

RESULT.   Let X be a random variable.   Then

$$V(X) = E(X^2) - (E(X))^2.$$

Proof.   From (*), $V(X) = E((X - E(X))^2)$
$= E(X^2 - 2E(X)X + (E(X))^2)$
$= E(X^2) - 2E(X).E(X) + (E(X))^2$
(using §56 treating E(X) as a constant)
$= E(X^2) - (E(X))^2.$

60.   EXAMPLE.   The continuous random variable X has p.d.f. given by

$$f(x) = \begin{cases} 4x(1 - x^2) & (0 \leq x \leq 1), \\ 0 & \text{otherwise.} \end{cases}$$

Find E(X),   V(X)   and the standard deviation of X.

Solution. $E(X) = \int_0^1 4x^2(1 - x^2)\ dx = \int_0^1 4x^2 - 4x^4\ dx$

$$= \left[\frac{4x^3}{3} - \frac{4x^5}{5}\right]_0^1 = \frac{4}{3} - \frac{4}{5} = \frac{8}{15}.$$

$$E(X^2) = \int_0^1 4x^3(1-x^2)\,dx = \left[x^4 - \frac{2}{3}x^6\right]_0^1 = \frac{1}{3}.$$

So, by §59, $V(X) = E(X^2) - (E(X))^2 = \frac{1}{3} - \frac{64}{225} = \frac{11}{225}$.

So the standard deviation $(\sigma_X) = \frac{\sqrt{11}}{15} = 0.221$.

EXAMPLES TO DO: Pages 64 - 67:   Exs. 3, 4, 13 - 16.

61. EXAMPLE. The random variable X has p.d.f. given by

$$f(x) = \begin{cases} 2xe^{-x^2} & (x \geq 0), \\ 0 & \text{otherwise.} \end{cases}$$

Find E(X) and V(X).

Solution. $E(X) = \int_0^\infty 2x^2 e^{-x^2}\,dx.$

Put $u = x^2$ in this, i.e. $x = u^{\frac{1}{2}}$.   So $dx = \frac{1}{2}u^{-\frac{1}{2}}du$.

So $E(X) = \int_0^\infty 2u \cdot e^{-u} \cdot \frac{1}{2}u^{-\frac{1}{2}}\,du = \int_0^\infty u^{\frac{1}{2}} e^{-u}\,du$

$$= \Gamma(3/2) = \tfrac{1}{2}\sqrt{\pi} = 0.886. \quad \text{(See below.)}$$

To find V(X), we must first find $E(X^2)$. So

$$E(X^2) = \int_0^\infty 2x^3 e^{-x^2}\,dx \quad (\text{Set } u = x^2.)$$

$$= \int_0^\infty 2u^{3/2} e^{-u} \cdot \tfrac{1}{2}u^{-\frac{1}{2}}\,du = \int_0^\infty u e^{-u}\,du$$

$$= \Gamma(2) = 1.$$

So we have, by §59, $V(X) = 1 - \frac{\pi}{4} = \frac{\pi - 4}{4} = 0.215$.

Notice: See the appendix (§100) for some notes on the Gamma function integral, which we have used in this example. Admittedly the integral for $E(X^2)$ here could be done by integration by parts, but be aware that integration by parts will not succeed on the integral for E(X).

EXAMPLES TO DO: Page 67:   Ex. 12.

62. HEREDITARY RESULTS FOR VARIANCE

The following results give the variance of some random variables related to given random variables. Notice part (ii) of Result 1 in particular.

RESULT 1. Let X, Y be random variables and let c be a constant. Then

(i) $V(X + c) = V(X) + c$;

(ii) $V(cX) = c^2 V(X)$.

Also, <u>provided that X and Y are independent</u>,

(iii) $V(X + Y) = V(X) + V(Y)$.

Proof. (i) Clearly, adding c to X does not alter the shape of the distribution. So it seems reasonable that the variance is unaltered.

(ii) $V(cX) = E(c^2 X^2) - (E(cX))^2$
$= c^2 (E(X^2) - (E(X))^2) = c^2 V(X)$.

(iii) See more advanced books.

RESULT 2. Let $X_1, X_2, \ldots, X_n$ be mutually independent random variables. Then

$$V(X_1 + X_2 + \ldots + X_n) = V(X_1) + V(X_2) + \ldots + V(X_n).$$

63. EXPECTATION AND VARIANCE FOR A BINOMIAL RANDOM VARIABLE

Suppose we toss a dice 72 times. We would expect about 12 sixes. The following result makes these feelings precise.

RESULT. Let X be a binomial random variable based on n repetitions and constant probability p of success. Then
$$E(X) = np \quad \text{and} \quad V(X) = np(1-p).$$

Proof. For each i with $1 \leq i \leq n$, define a random variable X_i by

$$X_i = \begin{cases} 1 & \text{if there is success on the ith repetition,} \\ 0 & \text{if there is failure on the ith repetition.} \end{cases}$$

So we have that X (as the total number of successes) is

given by $X = X_1 + X_2 + \ldots + X_n$ and the random variables X_i $(1 \leq i \leq n)$ are mutually independent. Now, from §48,

$$E(X_i) = 1.p + 0.(1-p) = p$$

and from §52(a),

$$E(X_i^2) = 1^2.p + 0^2.(1-p) = p.$$

So, by §59,

$$V(X_i) = E(X_i^2) - (E(X_i))^2 = p - p^2 = p(1-p).$$

Then using hereditary properties from §56 and §62, we see

$$E(X) = E(X_1) + E(X_2) + \ldots + E(X_n) = np$$

and

$$V(X) = V(X_1) + V(X_2) + \ldots + V(X_n) = np(1-p)$$

as required.

64. EXAMPLE. A dice is tossed 72 times. Let X be the number of sixes. Find $E(X)$ and $V(X)$, and the standard deviation of X.

Solution. Use the result of §63 with $n = 72$ and $p = \frac{1}{6}$. So

$$E(X) = 72 \cdot \frac{1}{6} = 12$$

and

$$V(X) = np(1-p) = 72 \cdot \frac{1}{6} \cdot \frac{5}{6} = 10.$$

So the standard deviation is $\sqrt{10} = 3.162$.

EXAMPLES TO DO: Page 67: Ex. 18.

65. THE AVERAGE OF N OBSERVATIONS

The following result lends support to the feeling that the average of n observations of a random variable is (in some sense) better than one single observation.

RESULT. Suppose that $X_1, X_2, \ldots, X_n$ are n independent observations of a random variable X with $E(X) = \mu$ and $V(X) = \sigma^2$. Let the random variable $\overline{X}$ be defined by

$$\overline{X} = \frac{1}{n}\left(X_1 + X_2 + \ldots + X_n\right).$$

Then $E(\overline{X}) = \mu$ and $V(\overline{X}) = \sigma^2/n$.

Proof.   We use the hereditary results of §56 and §62.

$$E(\bar{X}) = \frac{1}{n}\left[E(X_1) + E(X_2) + \ldots + E(X_n)\right] = \frac{1}{n} \cdot n\mu = \mu.$$

$$V(\bar{X}) = \frac{1}{n^2}\left[V(X_1) + V(X_2) + \ldots + V(X_n)\right] = \frac{1}{n^2} \cdot n\sigma^2 = \frac{\sigma^2}{n}.$$

Notice: The standard deviation for $\bar{X}$ is $\sigma/\sqrt{n}$, which is (for $n \geq 2$) smaller than the original standard deviation σ for X. This means that for example if we take the average of 25 observations of X to produce $\bar{X}$ then the standard deviation of $\bar{X}$ is $\sigma/5$ compared with σ for one single observation of X. As such $\bar{X}$ is much more likely to be close to the expected value μ than any one single observation.

EXAMPLES 8

1. A moderate darts player has the choice of trying to score (a) 60 (i.e. a treble 20), or (b) 42 (i.e. a treble 14), with each dart he throws. The diagrams below show the scores in the nine beds of the board in the vicinities of these trebles. He has probability 0.1 of hitting each bed and also probability 0.1 of bouncing off and scoring zero.

(a)

5	20	1
15	60	3
5	20	1

(b)

9	27	9
14	42	14
11	33	11

Find his expected score from one dart in each case.

2. (The dartboard) A standard dartboard is a disc consisting of a small central disc scoring 50, a narrow surrounding ring scoring 25, and then 20 equal sectors radiating out from the centre. The sectors are numbered from 1 to 20 in the order 20,1,18,4,13,6,10,15,2,17, 3,19,7,16,8,11,14,9,12,5 starting at the top and reading clockwise. Each sector is subdivided into 4 parts: the two larger parts score the sector number, while the outer small part scores double the sector number and the inner one scores treble the sector number. For example, here is a sketch of the sector numbered 6:

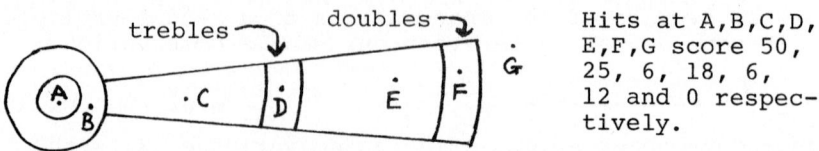

Hits at A,B,C,D, E,F,G score 50, 25, 6, 18, 6, 12 and 0 respectively.

An automaton throws darts at a dartboard. His distribution of aim is uniform over the scoring part of the board and his darts never bounce off. Show by a crude calculation ignoring doubles, trebles, the 25 and 50 that his expected score from three darts is at least 31.5.

3. Each of the following probability density functions corresponds to a continuous random variable X. (In each case we have only stated the definition where the p.d.f. is non-zero.) For each random variable X, find $E(X)$ and $V(X)$.

(a) $f(x) = 2x$ $\quad (0 \leq x \leq 1)$;

(b) $f(x) = 6x(1 - x)$ $\quad (0 \leq x \leq 1)$;

(c) $f(x) = 12x^2(1 - x)$ $\quad (0 \leq x \leq 1)$;

(d) $f(x) = \dfrac{1}{b - a}$ $\quad (a \leq x \leq b)$ (The uniform distribution §46);

(e) $f(x) = ae^{-ax}$ $\quad (x > 0)$ (a is a positive constant)

(The exponential distribution §45);

(f) $f(x) = a^2 x e^{-ax}$ $\quad (x > 0)$ (a is a positive constant);

4. A random variable X is uniformly distributed on [0, 4]. Find $E(X)$ and the standard deviation of X. Find also $E(X^3)$. Is it true that $E(X^3) = (E(X))^3$? Would you have expected this result and why?

5. A TV game show requires contestants to make up the longest word they can out of 10 given letters in a given time. For a word of x letters they win £x^2. (So a 4 letter word wins £16.) A man knows that for any given set of letters the probability that the length of his word is given by the following table:

Word length	1	2	3	4	5	6	7	8	9	10
Probability	0	0	.05	.05	.20	.30	.20	.10	.07	.03

Find his expected win on one word.

6. The life length of a heater element is T years where T is a random variable with the p.d.f.

$$f(t) = \begin{cases} \dfrac{1}{(t + 1)^2} & (t \geq 0), \\ 0 & \text{otherwise.} \end{cases}$$

It costs £A to manufacture an element and the manufacturer sells it for £B but gives a complete refund if it lasts less than 6 months. Find the manufacturer's expected profit on a heater element.

7. A disc of radius 3 units is used as a target for rifle shots. The random variable X is the distance from the centre of the target at which a shot strikes the plane of the target and X has p.d.f. given by

$$f(x) = \begin{cases} \frac{1}{3}e^{-x/3} & (x \geq 0), \\ 0 & \text{otherwise.} \end{cases}$$

On the target are marked circles with centres at the centre of the target and of radii 1 unit and 2 units. The scoring is 5 points for a hit in the central circular region, 2 points for a hit in the inner ring, 1 point for a hit in the outer ring and nothing for a shot which misses the target. Find the expected score from one shot.

8. The quality of a mass produced article is measured by a random variable X with p.d.f. given by

$$f(x) = \begin{cases} e^{-x} & (x > 0), \\ 0 & \text{otherwise.} \end{cases}$$

The net profit in selling the article is £2 when $0 < X < 1$, £3 when $1 \leq X \leq 2$ and £0.50 when $X > 2$. Find the mean profit per article.

9. The life length of an electrical item can be regarded as a continuous random variable X years with p.d.f. given by

$$f(x) = \begin{cases} \frac{1}{20}(2x+1) & (0 < x \leq 4), \\ 0 & \text{otherwise.} \end{cases}$$

It costs the manufacturer £A to make each item and he sells it for £B. He also guarantees a partial refund of £C on items with life length less than 1 year. He knows however that only D% of owners of items with life length less than 1 year actually claim their refund. Find his expected profit on an item.

10. The life length of a television set is T years, where T is a continuous random variable with p.d.f. f given by

$$f(t) = \begin{cases} \frac{8}{t^3} & (t \geq 2), \\ 0 & \text{otherwise.} \end{cases}$$

Note that this means every set lasts at least two years. A television dealer buys a set for £A and rents it out for the whole of its lifetime. He receives a rental of £B per year during the first three years of its life and £C per year after that. (The rental for an uncompleted year is in proportion.) The total repair costs to the dealer over the whole life of a set are £R, where R is uniformly distributed between 0 and 100. Find the dealer's expected profit on a set.

11. A TV dealer buys television sets for £A each and sells them for £B each. The life length of the sets is a random variable X years with the probability density function

$$f(x) = \begin{cases} \tfrac{1}{2}e^{-\tfrac{1}{2}x} & (x \geq 0), \\ 0 & \text{otherwise.} \end{cases}$$

The dealer gives a complete refund of £B on sets with life length less than 1 year and he gives a partial refund of £½B if the life length is between 1 and 2 years. Find his expected profit on a set.

He sells a set on a certain day and one year later he finds he has had no claim for a refund on the set. What is then the probability that a partial refund of £½B will be claimed on the set in the coming year?

12. Random variables X and Y have respective probability density functions f and g given by

$$f(x) = \begin{cases} 4xe^{-2x} & (x > 0), \\ 0 & \text{otherwise,} \end{cases} \quad g(x) = \begin{cases} 2y^3 e^{-y^2} & (y > 0), \\ 0 & \text{otherwise.} \end{cases}$$

Using the Gamma function, find $E(X)$, $V(X)$, $E(Y)$, $V(Y)$.

13. A normal coin is tossed 400 times. Let X be the number of heads. Find $E(X)$ and the standard deviation of X.

14. For the darts player in Ex.1 above, find the standard deviation of the score from one dart in each of the cases (a) and (b).

15. Find the standard deviation of the money won on each word by the contestant in Ex.5 above.

16. Two dice are thrown together and the total score X is noted. (So $2 \leq X \leq 12$.) Find $E(X)$ and the standard deviation of X.

17. A continuous random variable X takes all non-negative real values and has the p.d.f.

$$f(x) = \frac{2}{\pi(1 + x^2)} \quad (0 \leq x < \infty).$$

Check that this is a valid probability density function. Show that the integrals for $E(X)$ and $E(X^2)$ do **not** converge.

18. From an ordinary pack of 52 cards, one card is chosen at random. This experiment is done 300 times and the total number of hearts obtained gives a random variable X. Find $E(X)$ and the standard deviation of X.

IX. THE POISSON DISTRIBUTION

66. THE BACKGROUND

Think for example of the random variables given by:

1. The number of thunderstorms per year in a town.
2. The number of accidents in a factory per week.
3. The number of customers arriving at a shop in a five minute interval.
4. The number of supertankers arriving at an oil terminal in a week.
5. The number of errors on one page of a book.

These are all discrete random variables taking the possible values 0, 1, 2, 3, 4, ... , with the small values being the most likely. In fact these could be typical Poisson random variables. We shall say more about how they arise in §75.

67. THE POISSON DISTRIBUTION; POISSON RANDOM VARIABLES

Definition. Let X be a discrete random variable taking the values 0, 1, 2, 3, ... , and such that

$$P(X = k) = \frac{e^{-\lambda}\lambda^k}{k!} \quad (k = 0, 1, 2, 3, \ldots).$$

Then X is said to have a <u>Poisson distribution with parameter</u> λ (or to be a <u>Poisson random variable with parameter</u> λ).

<u>Notice</u>: (i) Here λ is a positive constant.

(ii) While the binomial distribution (§36) has two controlling parameters n and p, the Poisson distribution has only one, namely λ. The distribution of probability among the values 0, 1, 2, 3, ... is fixed immediately λ is known.

(iii) It is a routine exercise in Analysis to show that, for every value of λ,

$$\frac{\lambda^k}{k!} \to 0 \quad \text{as} \quad k \to \infty.$$

Consequently the amounts of probability assigned to successive values of k eventually tail off. For example, for the case of $\lambda = 3$, the distribution of probability given by

$$P(X = k) = \frac{e^{-3}3^k}{k!}$$

is shown in the following table:

k	P(X = k)	k	P(X = k)
0	0.050	5	0.101
1	0.149	6	0.050
2	0.224	7	0.022
3	0.224	8	0.008
4	0.168	9	0.003

68. EXPECTATION AND VARIANCE FOR THE POISSON DISTRIBUTION

RESULT. Let X be a Poisson random random variable with parameter λ. Then

$$E(X) = \lambda \quad \text{and} \quad V(X) = \lambda.$$

Proof. From the definition of §48, notice first that

$$E(X) = \sum_{k=0}^{\infty} k \cdot \frac{e^{-\lambda}\lambda^k}{k!}. \qquad \ldots (1)$$

In an attempt to build up (1), we start from the series for the exponential function

$$e^{\lambda} = \sum_{k=0}^{\infty} \frac{\lambda^k}{k!}.$$

Differentiate this with respect to λ. This gives

$$e^{\lambda} = \sum_{k=0}^{\infty} k \frac{\lambda^{k-1}}{k!}.$$

Now multiply by λ to obtain

$$\lambda e^{\lambda} = \sum_{k=0}^{\infty} k \frac{\lambda^k}{k!}. \qquad \ldots (2)$$

Now divide through by e^{λ}. This gives

$$\lambda = \sum_{k=0}^{\infty} k \cdot \frac{e^{-\lambda}\lambda^k}{k!} = E(X) \quad \text{(as in (1))},$$

i.e. $E(X) = \lambda$.

Now differentiate (2) with respect to λ. This gives

$$\lambda e^{\lambda} + e^{\lambda} = \sum_{k=0}^{\infty} k^2 \frac{\lambda^{k-1}}{k!}.$$

Now multiply through by $\lambda e^{-\lambda}$.   This gives

$$\lambda^2 + \lambda = \sum_{k=0}^{\infty} k^2 \frac{e^{-\lambda}\lambda^k}{k!} = E(X^2).$$

So $V(X) = E(X^2) - (E(X))^2 = \lambda^2 + \lambda - \lambda^2 = \lambda.$

Notice: (i) In observing a Poisson experiment you may not know the value of λ. For example if you are observing the number of supertankers arriving at an oil terminal per week (which you suspect to be a Poisson random variable) and 156 supertankers arrive in 52 weeks then you could reasonably estimate that the value of $\lambda = 156/52 = 3$, on the basis that $E(X) = \lambda$. A full discussion of the validity and accuracy of such estimates lies beyond the scope of this book: the theory of estimation is a large subject in its own right.

(ii) In the above proof we used without comment the fact that it is permissible to differentiate a power series term by term inside its circle of convergence. Books on Analysis give justification of this manoeuvre.

69.   **EXAMPLE.** The number of ore carriers arriving at a deep water terminal per week is a Poisson random variable with parameter 2. Find the probability that in a given week there are

(a) exactly three ore carriers,
(b) at least two ore carriers,
(c) at most four given that there are at least two.

Solution. Here we have $\lambda = 2$. So

(a) $P(X = 3) = \dfrac{e^{-2} 2^3}{3!} = \dfrac{4}{3} e^{-2} = 0.180.$

(b) $P(X \geq 2) = P(X=2) + P(X=3) + P(X=4) + \ldots$

$\qquad\qquad = 1 - P(X=0) - P(X=1)$

$\qquad\qquad = 1 - e^{-2} - 2e^{-2} = 1 - 3e^{-2} = 0.594.$

(c) $P(X \leq 4 \mid X \geq 2) = \dfrac{P((X \leq 4) \cap (X \geq 2))}{P(X \geq 2)}$ (by §18(ii))

$\qquad\qquad = \dfrac{P(2 \leq X \leq 4)}{P(X \geq 2)}$

$\qquad\qquad = \dfrac{P(X=2) + P(X=3) + P(X=4)}{P(X \geq 2)}$

$\qquad\qquad = \dfrac{2e^{-2} + \frac{4}{3}e^{-2} + \frac{2}{3}e^{-2}}{1 - 3e^{-2}} = 0.911.$

Notice: In (b) we avoid summing an infinite series of terms by looking at the complementary event and subtracting the probability of that event from 1.

EXAMPLES TO DO: Pages 76 - 77: Exs. 1, 2(a),(b),(d).

70. POISSON APPROXIMATION TO THE BINOMIAL - THE BACKGROUND

Suppose that we are searching for top-flight pianists. We may know that the probability that a person chosen at random from the population is a top-flight pianist is 0.001. We may then ask a question like:

What is the probability that in a group of 4000 randomly chosen people there are exactly 3 top-flight pianists?

This problem is really binomial with n = 4000 and p = 0.001. So if we let X be the number of top-flight pianists, the question can be answered (using the formula from §37) as:

$$P(X=3) = \binom{4000}{3}(.001)^3(.999)^{3997}. \quad \ldots \quad (*)$$

This involves some hefty arithmetic - something which did not show up to any great extent in earlier examples on the binomial distribution (§36 - §40). It is really the fact that n is "large" and p is "small" (i.e. n = 4000, p = 0.001) that causes the trouble. Fortunately in such cases we can avoid using formula (*) and use instead the Poisson approximation to the binomial: we give the theory in §71 and examples in §72 - §73.

71. POISSON APPROXIMATION TO THE BINOMIAL - THE THEORY

In general terms, the following result says that when n is large and p is small (see (ii) below), the binomial distribution with parameters n and p approaches the Poisson distribution with parameter given by $\lambda = np$.

RESULT. Let X be a binomial random variable with n repetitions and constant probability p of success, so that

$$P(X=k) = \binom{n}{k}p^k(1-p)^{n-k}.$$

Now let $n \to \infty$ and $p \to 0$ in such a way that $np = \lambda$, where λ is a constant. Then

$$P(X=k) \to \frac{e^{-\lambda}\lambda^k}{k!} \quad \text{as } n \to \infty \ \& \ p \to 0 \text{ with } np = \lambda.$$

Proof. We are addressing all cases where $np = \lambda$, where λ has a constant value. So $p = \lambda/n$ and once λ is fixed, we are looking at the behaviour of $P(X = k)$ as $n \to \infty$. So translating p as λ/n gives

$$P(X = k) = \binom{n}{k} p^k (1-p)^{n-k} = \binom{n}{k} \left(\frac{\lambda}{n}\right)^k \left(1 - \frac{\lambda}{n}\right)^{n-k}$$

$$= \frac{n(n-1)(n-2) \cdots (n-k+1)}{k!} \cdot \frac{\lambda^k}{n^k} \cdot \frac{\left(1 - \frac{\lambda}{n}\right)^n}{\left(1 - \frac{\lambda}{n}\right)^k}$$

$$= \frac{\lambda^k}{k!} \cdot \frac{1 \cdot \left(1 - \frac{1}{n}\right)\left(1 - \frac{2}{n}\right) \cdots \left(1 - \frac{k-1}{n}\right)}{\left(1 - \frac{\lambda}{n}\right)^k} \left(1 - \frac{\lambda}{n}\right)^n \quad \ldots (*)$$

$$\to \frac{\lambda^k}{k!} \cdot \frac{1 \cdot 1 \cdot 1 \cdots 1}{1} \cdot e^{-\lambda} \quad \text{as } n \to \infty. \quad \begin{pmatrix} \text{See (i)} \\ \text{below.} \end{pmatrix}$$

$$= \frac{e^{-\lambda} \lambda^k}{k!} \cdot \qquad \ldots (**)$$

Notice: (i) The proof is fairly delicate though the result is fairly straightforward. In line (*) keep in mind that we have fixed λ and we are working out $P(X = k)$ for some particular value of k which is by this stage chosen (and so is fixed). This means that by taking n sufficiently large we can make all the factors

$$\left(1 - \frac{1}{n}\right), \left(1 - \frac{2}{n}\right), \ldots, \left(1 - \frac{k-1}{n}\right) \text{ and } \left(1 - \frac{\lambda}{n}\right)^k$$

as close to 1 as we like. In other words each of these factors tends to 1 as $n \to \infty$.

Also notice in line (*) that there is a standard result in Analysis which says that, for every real number x, we have that

$$\left(1 + \frac{x}{n}\right)^n \to e^x \quad \text{as } n \to \infty.$$

It is this result which gives the factor $e^{-\lambda}$.

(ii) In using the above result in problems, the crux of the matter is that we really want to replace the <u>binomial formula</u> for $P(X = k)$ by the <u>Poisson formula</u>, i.e. we want to replace line (*) in the above proof by line (**). Examining the proof shows that this replacement has hope of success if the following three conditions are met:

(a) The factors $\left(1 - \frac{1}{n}\right)$, $\left(1 - \frac{2}{n}\right)$, ... , $\left(1 - \frac{k-1}{n}\right)$
are all close to 1, and the fewer of them the better.

(b) $\left(1 - \frac{\lambda}{n}\right)^k$ is close to 1.

(c) $\left(1 - \frac{\lambda}{n}\right)^n$ is close to $e^{-\lambda}$.

We now look at the requirements that these conditions place on the values of n, p and k.

Firstly, we can translate (a) to say "k should not be too big (in fact, the smaller the better), and n should be much bigger than k". So, say we restrict to $0 \leq k \leq 10$ and n much bigger than 10. Secondly, it is clear that (b) will be seriously prejudiced (for $k \geq 1$) if for example $\lambda/n = \frac{1}{2}$, (i.e. if $p = \frac{1}{2}$) because the factor then becomes $(\frac{1}{2})^k$, which is not close to 1. On the other hand, if $\lambda/n = 0.01$, (i.e. if $p = 0.01$) the factor is $(0.99)^k$, which is reasonably close to 1 for $0 \leq k \leq 10$. So, on this evidence, lower values of p, like 0.01, are much more likely to give a good approximation than higher values, like $p = \frac{1}{2}$. Thirdly, looking at (c) shows that, in general, the larger n is, the better the agreement between $\left(1 - \frac{\lambda}{n}\right)^n$ and $e^{-\lambda}$.

We therefore now make the statement that the binomial distribution can be approximated by the Poisson distribution when "n is large and p is small". The advantage is that the Poisson formula is generally easier to handle than the binomial one in these circumstances.

We really need to settle what we mean by "n large and p small". Generally speaking it is reasonable to limit the use of the Poisson approximation to cases where $n \geq 50$ and $p \leq 0.1$, though obviously it depends on how accurate you wish your answer to be. Some books are more conservative demanding that n should be larger and p smaller than these values. The example in §73 examines the agreement for certain values of n and p. It is also worth noting that if in addition to $n \geq 50$ and $p \leq 0.1$ we also have $np \geq 5$, then it may be worth considering the normal approximation (§96) rather than the Poisson approximation.

(iii) Recall that for a binomial random variable X with n repetitions and probability p of success $E(X) = np$, while for a Poisson random variable X with parameter λ, $E(X) = \lambda$. This lies behind the choice of $\lambda = np$ in the above result: it is reasonable that the expected value of the approximation should agree with that of the original.

72. EXAMPLE. The probability that a person chosen at random from a certain population belongs to a particular rare blood group is 0.002. Find the probability that in a group of 2000 randomly chosen people there are

(a) exactly three such people,
(b) at least two such people.

Solution. Let X be the number of people with that blood group.

Then X is binomial with $n = 2000$ and $p = 0.002$ and we can safely apply the Poisson approximation with $\lambda = 2000 \times 0.002 = 4$. So the answers are

(a) $P(X = 3) = \dfrac{e^{-4} 4^3}{3!} = \dfrac{32 e^{-4}}{3} = 0.195$.

(b) $P(X \geq 2) = 1 - P(X = 0) - P(X = 1)$

$= 1 - \dfrac{e^{-4} 4^0}{0!} - \dfrac{e^{-4} 4^1}{1!} = 1 - 5e^{-4}$

$= 0.908$.

EXAMPLES TO DO: Page 77: Ex. 4, 5, 6.

73. COMPARISON OF THE BINOMIAL AND THE POISSON APPROXIMATION TO IT

In §71 some comments were made about the accuracy of the Poisson approximation to the binomial in relation to the values of n and p. The following example takes this further.

EXAMPLE. Let X be a binomial random variable with parameters n and p. Draw tables comparing the true probabilities of various values of X (as given by the binomial) with the approximate probabilities (as given by the Poisson approximation). Consider the four cases

(a) $n = 1000$, $p = 0.002$,
(b) $n = 100$, $p = 0.02$,
(c) $n = 20$, $p = 0.1$,
(d) $n = 10$, $p = 0.2$.

Within each case the table is to show the values of $P(X = k)$ for $k = 0, 1, 2, \ldots, 9$.

Solution. In each of the four cases the Poisson approximation is given by $\lambda = 2$, since $\lambda = np$. So the Poisson approximation gives the same set of probabilities in each of the four cases. The following table therefore gives the value of X in column 1, the binomial probabilities for the cases (a), (b), (c), (d) in columns 2, 3, 4, 5 and then the estimate of the probability given by the Poisson approximation in column 6.

	(a) $n=1000$ $p=.002$	(b) $n=100$ $p=.02$	(c) $n=20$ $p=.1$	(d) $n=10$ $p=.2$	Poisson approximation
0	.1351	.1326	.1216	.1074	.1353
1	.2707	.2707	.2702	.2684	.2707
2	.2709	.2734	.2852	.3020	.2707
3	.1806	.1823	.1901	.2013	.1804
4	.0902	.0902	.0898	.0881	.0902
5	.0360	.0353	.0319	.0264	.0361
6	.0120	.0114	.0089	.0055	.0120
7	.0034	.0031	.0020	.0008	.0034
8	.0008	.0007	.0004	.0001	.0009
9	.0002	.0002	.0001	.0000	.0002

EXAMPLES TO DO: Page 77:   Ex. 3.

74.   THE SUM OF INDEPENDENT POISSON RANDOM VARIABLES

We state the following result without proof.

RESULT. Let $X_1, X_2, \ldots, X_n$ be n independent Poisson random variables with parameters $\lambda_1, \lambda_2, \ldots, \lambda_n$. Then $(X_1 + X_2 + \ldots + X_n)$ is a Poisson random variable with parameter $(\lambda_1 + \lambda_2 + \ldots + \lambda_n)$.

Proof. See more advanced books.

The following example illustrates how the above result can be used.

EXAMPLE. The number of mechanical breakdowns in a certain factory per day can be regarded as a Poisson random variable with parameter 1.5. Find the probability that
(a) there will be exactly three breakdowns on a certain day,
(b) there will be exactly six breakdowns in a five day period.

Solution. (a) Let X be the number of breakdowns in a day.

Then $P(X = 3) = \dfrac{e^{-1.5}(1.5)^3}{3!} = 0.126$.

(b) Let Y be the total number of breakdowns in a five day period.

We can regard Y as the sum of 5 independent Poisson random variables each with parameter 1.5. So by the abve result, Y is a Poisson random variable with parameter 7.5. Hence the required answer is

$$P(Y = 6) = \dfrac{e^{-7.5}(7.5)^6}{6!} = 0.137.$$

EXAMPLES TO DO: Page 77: Exs. 2(c), 7, 8.

75.   Many would be Poisson random variables really arise as the Poisson approximation to the binomial in cases when n is large and p is small. If you look at the examples in §66 you can probably see this yourself.

For example, the number of thunderstorms per year in Edinburgh might be thought of as a binomial random variable, with n = 365 and p = the small probability that on any one day there will be a thunderstorm. (There may be a question mark over the independence from day to day and no doubt it can be shown with statistics that p is not constant throughout the year, but don't let such considerations veil your appreciation of the main issue.)

EXAMPLES 9

1. The number of street musicians seen by a lady on her weekly visit to town is a Poisson random variable with parameter 3. Find the probability that on a given day she sees

   (a) no street musician,
   (b) at least three street musicians,
   (c) at most three given that she sees at least one.

2. The number of breakdowns of a machine on any particular working day can be regarded as a Poisson random variable with parameter 2. Find the probability that

   (a) on a given working day there will be exactly three breakdowns,
   (b) on a given working day there will be at least three breakdowns,

(c) during a period of five working days there will be exactly two breakdowns,
(d) during a period of five working days there will be exactly four working days without a breakdown.

3. (Comparison of binomial and Poisson approximation) The probability that a newly delivered bus will have to be sent back to the makers with defects is 0.1. The condition of each bus can be taken as independent of the others. Use (a) the binomial, (b) the Poisson approximation, to find the probability that in a batch of 10 newly delivered buses exactly 0, 1, 2, 3 require to be sent back. Compare the answers from the two methods.

4. The probability that a person has a certain tissue type is 0.02. Using the Poisson distribution, calculate the probability that in a random sample of 150 people

(a) exactly two have this tissue type,
(b) at least two have this tissue type.

If the 150 people are tested one after another in random order and exactly two people with the required tissue type are found in the first thirty people tested, what is the probability that at least two more such people will be found among the remaining 120 people?

5. The probability that a person chosen at random from the population is suitable for training as a special agent is 0.0006. Use the Poisson approximation to the binomial to find the probability that a group of 10000 people contains

(a) exactly two suitable people,
(b) at least four suitable people,
(c) at least four given that there are at most six.

6. The probability that a railway hopper wagon will suffer a hot axle box on a certain trip is 0.001. If 1000 such wagons are making the trip every day, find the probability that at least one wagon will suffer a hot axle box on a given day. Trains are made up of 40 such wagons coupled together. Find the probability that such a train will make the trip without suffering a hot axle box. If a train driver makes 100 such trips (each with 40 wagons) find the probability that at least three trips will give rise to a hot axle box.

7. The lady in Ex. 1 makes three visits to town. Find the probability that over the three visits she sees

(a) a total of exactly three street musicians,
(b) a total of less than three street musicians.

8. The lady in Ex. 1 gives money to the street musicians she sees under two possible schemes:
- (a) £1 to each street musician she sees;
- (b) £1 to the first, £2 to the second, £3 to the third and so on.

Find the expected total amount of money she gives out in a day under each scheme.

X. THE NORMAL DISTRIBUTION

76. THE NORMAL DISTRIBUTION

This is a very common distribution which is of both practical and theoretical importance.

<u>Definition.</u> Let X be a continuous random variable taking all real values in the range $(-\infty, \infty)$ and with p.d.f. given by

$$f(x) = \frac{1}{\sigma \sqrt{(2\pi)}} e^{-\frac{1}{2}((x-\mu)/\sigma)^2} \quad (-\infty < x < \infty).$$

Then X is said to have a <u>normal distribution</u> with parameters μ and σ (or to be a <u>normal random variable</u> with parameters μ and σ). We write X is $N(\mu, \sigma^2)$.

<u>Notice</u>: (i) Here the numbers μ and σ are constants (with σ positive).

(ii) It turns out (see §83) that

$$E(X) = \mu \quad \text{and} \quad V(X) = \sigma^2,$$

so that μ is the <u>mean</u> and σ is the <u>standard deviation</u> for the distribution.

(iii) The distribution is symmetrical about its mean μ. Altering μ shifts the distribution along the x-axis. Altering σ (the standard deviation) alters the shape (peakedness) of the distribution. Here is a rough sketch of the shape of the p.d.f.:

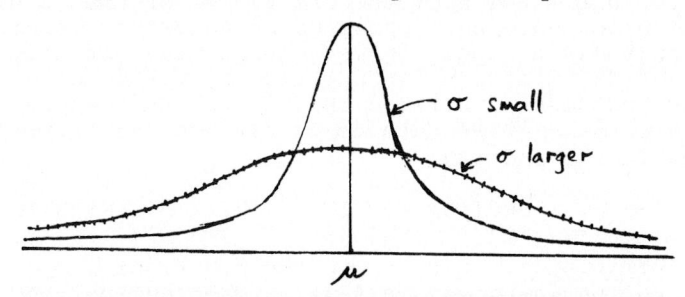

(iv) As you might expect from the shape of the p.d.f., many distributions arising in practice have normal or roughly normal characteristics and so can be approximated by the $N(\mu, \sigma^2)$ distribution for suitable choices of the parameters μ and σ.

77. THE STANDARD NORMAL DISTRIBUTION $N(0,1)$

The simplest case for the normal distribution is the case with mean zero and standard deviaton one, i.e. the $N(0,1)$ case. This is the so called <u>standard normal distribution</u>.

<u>Definition</u>. Let X be a continuous random variable taking all real values in the range $(-\infty, \infty)$ and with p.d.f.

$$\phi(x) = \frac{1}{\sqrt{(2\pi)}} e^{-\frac{1}{2}x^2} \qquad (-\infty < x < \infty).$$

Then X is said to have the <u>standard normal distribution</u> $N(0,1)$.

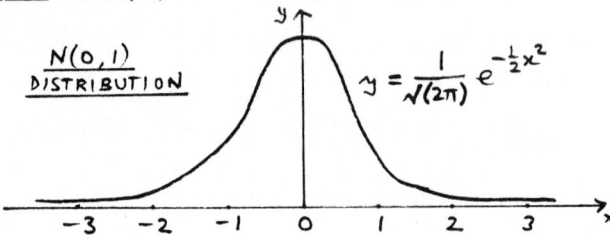

<u>Notice</u>: Recall from §41 that for a continuous distribution

$$P(a \leq X \leq b) = \int_a^b f(x)\, dx,$$

where f is the p.d.f.. This was used in the example in §42. Unfortunately the integration involved in the evaluation of such integrals for the normal distribution cannot be done in terms of elementary functions. To resolve this, tables of the c.d.f. Φ of the $N(0,1)$ distribution have been produced by numerical methods (see page 118). These tables can be used not only for the $N(0,1)$ distribution but also by virtue of a scaling process (described in §80) for the general case of $N(\mu,\sigma^2)$. More details of Φ and its properties follow in §78 and §79.

78. THE Φ FUNCTION FOR THE $N(0,1)$ DISTRIBUTION

<u>Definition</u>. Let ϕ be the p.d.f. of the $N(0,1)$ distribution (as given in §77). Then define the function Φ by

$$\Phi(a) = \int_{-\infty}^{a} \phi(x)\, dx$$

for each real number a.

<u>Notice</u>: (i) Φ is the cumulative distribution function (c.d.f.) for the $N(0,1)$ distribution. So we can picture $\Phi(a)$ as in the diagram.

$\Phi(a)$ is the shaded area

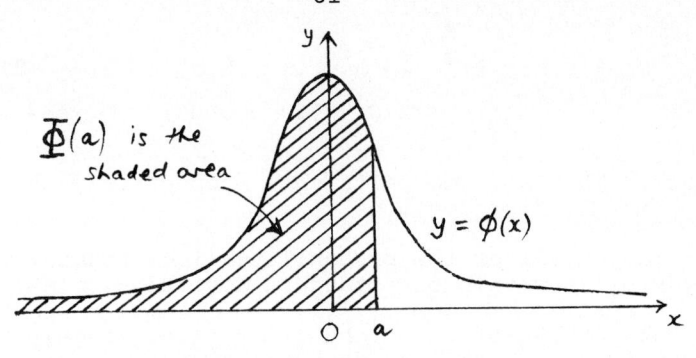

$y = \phi(x)$

(ii) A table of the function Φ is given on page 118. Here is an extract.

a	$\Phi(a)$
0	0.5000
0.5	0.6915
1.0	0.8413
1.5	0.9332
2.0	0.9772
2.5	0.9938
3.0	0.9987

Notice that, because the distribution is symmetrical about 0, half the "probability sand" lies to the left of 0 and half to the right. Notice also that Φ increases from 0 up to 1 as x runs from $-\infty$ up to $+\infty$. The table is only given for $a \geq 0$ since the values for $a < 0$ can be deduced by symmetry.

79. PROPERTIES OF THE Φ FUNCTION

RESULT. Let X be an $N(0,1)$ random variable, and let a and b be constants. Then

(i) $P(X \leq a) = \Phi(a)$; (ii) $P(X \geq a) = 1 - \Phi(a)$;
(iii) $P(a \leq X \leq b) = \Phi(b) - \Phi(a)$; (iv) $\Phi(0) = \frac{1}{2}$;
(v) $P(-a \leq X \leq a) = 2P(0 \leq X \leq a) = 2(\Phi(a) - \frac{1}{2})$;
(vi) $\Phi(-a) = 1 - \Phi(a)$.

Proof. Easy from the definition of Φ in §78.

EXAMPLE. Let X be an $N(0,1)$ random variable. Find $P(X \leq 1)$, $P(0.5 \leq X < 1.5)$, $P(-1 \leq X \leq 1)$, $P(-2 \leq X \leq 2)$.

Solution. $P(X \leq 1) = \Phi(1) = 0.8413$.

$P(0.5 \leq X \leq 1.5) = \Phi(1.5) - \Phi(0.5) = 0.9332 - 0.6915 = 0.2417$.

$P(-1 \leq X \leq 1) = 2P(0 \leq X \leq 1) = 2(\Phi(1) - \Phi(0))$
$= 2(0.8413 - 0.5000) = 0.6826$.

$$P(-2 \leq X \leq 2) = 2 P(0 \leq X \leq 2) = 2(\Phi(2) - \Phi(0))$$
$$= 2(0.9772 - 0.5000) = 0.9544.$$

Remark. The last two parts show respectively that for the $N(0,1)$ distribution

(a) 68.3% of the probability (i.e. roughly two thirds) lies within <u>one</u> standard deviation of the mean.

(b) 95.4% of the probability (i.e. roughly 95%) lies within <u>two</u> standard deviations of the mean.

We return to this remark later.

80. REDUCTION OF THE GENERAL CASE $N(\mu,\sigma^2)$ TO $N(0,1)$

The following result gives the connection.

RESULT. Let X be $N(\mu,\sigma^2)$. Let $Y = \dfrac{X - \mu}{\sigma}$. Then Y is $N(0,1)$.

Proof. Let X have p.d.f. f and c.d.f. F.
Let Y have p.d.f. g and c.d.f. G.

Then $G(a) = P(Y \leq a) = P\left(\dfrac{X - \mu}{\sigma} \leq a\right)$

$$= P(X \leq \mu + \sigma a)$$
$$= F(\mu + \sigma a).$$

Now differentiate both sides w.r.t. a remembering that the p.d.f. is the derivative of the c.d.f.. So

$$G'(a) = \sigma F'(\mu + \sigma a),$$

i.e. $g(a) = \sigma f(\mu + \sigma a)$

$$= \sigma \cdot \frac{1}{\sigma \sqrt{(2\pi)}} e^{-\frac{1}{2}((\mu + \sigma a - \mu)/\sigma)^2}$$

(on using the p.d.f. for the $N(\mu,\sigma^2)$)

$$= \frac{1}{\sqrt{(2\pi)}} e^{-\frac{1}{2}a^2}, \quad \text{i.e. the p.d.f. for the } N(0,1).$$

So g is the p.d.f. for the $N(0,1)$ distribution. So we conclude that Y is $N(0,1)$.

Notice: The examples of §81 and §82 show how to use this result.

81. EXAMPLE. Light bulbs have life length X hours, where X can be regarded as approximately an $N(1200, 2500)$ random variable. Find the probability that

(a) a bulb lasts more than 1230 hours,
(b) a bulb lasts between 1100 and 1300 hours,
(c) a bulb lasts less than 1250 hours given it lasts more than 1230 hours.

If three such bulbs are installed in a lift, what is the probability that at least one will still be working after a period of 1230 hours?

<u>Solution.</u> We reduce to the $N(0,1)$ case using §80. This gives $Y = \dfrac{X - 1200}{50}$ as $N(0,1)$. We then apply the tables from page 118 to Y as follows.

(a) $P(X > 1230) = P\left(\dfrac{X - 1200}{50} > \dfrac{1230 - 1200}{50}\right) = P(Y > 0.6)$

$= 1 - \Phi(0.6) = 1 - 0.7257 = 0.2743$.

(b) $P(1100 \leq X \leq 1300) = P\left(\dfrac{1100-1200}{50} \leq \dfrac{X-1200}{50} \leq \dfrac{1300-1200}{50}\right)$

$= P(-2 \leq Y \leq 2) = 2P(0 \leq Y \leq 2)$

$= 2(0.9772 - 0.5000) = 0.9544$.

(c) $P(X < 1250 \mid X > 1230) = \dfrac{P(1230 < X < 1250)}{P(X > 1230)}$ (by §18(ii))

$= \dfrac{P(0.6 < Y < 1)}{P(Y > 0.6)} = \dfrac{\Phi(1) - \Phi(0.6)}{1 - \Phi(0.6)}$

$= \dfrac{0.8413 - 0.7257}{1 - 0.7257} = \dfrac{1156}{2743} = 0.4214$.

For the <u>last part</u>, let Z be the number of bulbs still burning after a period of 1230 hours. This is <u>binomial</u> with $n = 3$ and $p = 0.2743$ (from part (a)). So,

the answer $= P(Z \geq 1) = P(Z = 1) + P(Z = 2) + P(Z = 3)$

$= 1 - P(Z = 0) = 1 - \binom{3}{0}(0.2743)^0(0.7257)^3$

$= 0.6178$.

<u>Remarks.</u> (i) In this question, the standard deviation is 50. So we can think of it as in the diagram overleaf which effectively counts off standard deviations on either side of the mean.

```
    |————|————|————|————|————|————→
   1100  1150 1200  1250 1300
              MEAN
    -2    -1    0    1    2
```

The numbers -2, -1, 0, 1, 2 etc give the number of standard deviations away from the mean. It is these numbers which are in effect the values of the $N(0,1)$ random variable Y.

(ii) We can now generalise the result in §79 to say that for <u>every</u> normal distribution (not just for $N(0,1)$)

(a) 68.3% of the probability (i.e. roughly two thirds) lies within <u>one</u> standard deviation of the mean.

(b) 95.4% of the probability (i.e. roughly 95%) lies within <u>two</u> standard deviations of the mean.

EXAMPLES TO DO: Page 87: Exs. 1 - 5.

82. <u>EXAMPLE</u>. Two mechanics A and B each perform the same type of standard repair on a car. The times taken by each mechanic are X_1 and X_2 and these can be regarded as approximately normal random variables with the distributions $N(96,225)$ and $N(100,400)$ respectively. Which mechanic is more likely to complete the repair if the time available is

(i) 100 minutes,
(ii) 75 minutes?

<u>Solution</u>. Here we reduce the performance of both mechanics to $N(0,1)$ (using §80) and thereby compare them. So we take

$$Y = \frac{X_1 - 96}{15} \text{ for A } \text{ and } Y = \frac{X_2 - 100}{20} \text{ for B.}$$

(i) $P(X_1 \leq 100) = P\left(Y \leq \frac{100 - 96}{15}\right) = \Phi(4/15)$, while

$P(X_2 \leq 100) = P\left(Y \leq \frac{100 - 100}{20}\right) = \Phi(0)$.

Now $\Phi(4/15) > \Phi(0)$. So mechanic A is more likely to complete it within 100 minute period.

(ii) $P(X_1 \leq 75) = P\left(Y \leq \frac{75 - 96}{15}\right) = \Phi(-1.4)$, while

$P(X_2 \leq 75) = P\left(Y \leq \frac{75 - 100}{20}\right) = \Phi(-1.25)$.

Now $\Phi(-1.25) > \Phi(-1.4)$. So mechanic B is more likely to complete the job within a 75 minute period.

Notice: You don't need the Φ-table on page 118 to do this type of example. You do however need to know that if $a < b$ then $\Phi(a) < \Phi(b)$ and to understand the significance of $\Phi(a)$. Treat such examples with considerable respect as it is easy to get things the wrong way round.

EXAMPLES TO DO: Pages 87-88: Exs. 6 - 9.

83. MEAN AND VARIANCE FOR THE NORMAL DISTRIBUTION

We prove here a result stated in §76.

RESULT. Let X be an $N(\mu, \sigma^2)$ random variable. Then
$$E(X) = \mu \quad \text{and} \quad V(X) = \sigma^2.$$

Proof. Firstly, $E(X) = \dfrac{1}{\sigma\sqrt{(2\pi)}} \displaystyle\int_{-\infty}^{\infty} x\, e^{-\frac{1}{2}((x-\mu)/\sigma)^2}\, dx$

$= \dfrac{1}{\sigma\sqrt{(2\pi)}} \left\{ \displaystyle\int_{-\infty}^{\infty} (x-\mu)\, e^{-\frac{1}{2}((x-\mu)/\sigma)^2}\, dx + \int_{-\infty}^{\infty} \mu\, e^{-\frac{1}{2}((x-\mu)/\sigma)^2}\, dx \right\}$

$= \dfrac{\sigma}{\sqrt{(2\pi)}} \displaystyle\int_{-\infty}^{\infty} y\, e^{-\frac{1}{2}y^2}\, dy + \dfrac{\mu}{\sigma\sqrt{(2\pi)}} \cdot \int_{-\infty}^{\infty} e^{-\frac{1}{2}((x-\mu)/\sigma)^2}\, dx$

on setting $y = (x-\mu)/\sigma$ in the first integral, so that $dx = \sigma\, dy$.

The first integral is then zero, because the integral involves as much area below the x-axis as it does above it. The second term has the value μ. To see this notice that it is really just μ times the integral of the p.d.f. of an $N(\mu,\sigma^2)$ random variable over $(-\infty, \infty)$ and such an integral has the value 1 as in (b) of §41(iii). So we concude that $E(X) = \mu$.

Secondly, $V(X) = E((X-\mu)^2)$ (by §58)

$= \dfrac{1}{\sigma\sqrt{(2\pi)}} \displaystyle\int_{-\infty}^{\infty} (x-\mu)^2\, e^{-\frac{1}{2}((x-\mu)/\sigma)^2}\, dx$

$= \dfrac{\sigma^2}{\sqrt{(2\pi)}} \displaystyle\int_{-\infty}^{\infty} y^2 e^{-\frac{1}{2}y^2}\, dy \quad \left[\text{setting } y = \dfrac{x-\mu}{\sigma}\right]$

$= \dfrac{2\sigma^2}{\sqrt{(2\pi)}} \displaystyle\int_{0}^{\infty} y^2 e^{-\frac{1}{2}y^2}\, dy \quad \text{(by symmetry)}$

$= \sigma^2$ (on using the result in §100 with $\alpha = 3$, $\beta = 2$ and $k = \frac{1}{2}$).

84. SUM OF INDEPENDENT NORMAL RANDOM VARIABLES

In §74 we saw that the sum of independent Poisson random variables is again Poisson. The following is the analogue for normal random variables.

RESULT. Let $X_1, X_2, \ldots, X_n$ be independent normal random variables where X_i has the distribution $N(\mu_i, \sigma_i^2)$. Then $(X_1 + X_2 + \ldots + X_n)$ is also a normal random variable and has the distribution

$$N(\mu_1 + \mu_2 + \ldots + \mu_n, \sigma_1^2 + \sigma_2^2 + \ldots + \sigma_n^2).$$

Proof. See more advanced books.

The following example illustrates a use of this result.

EXAMPLE. The heights of men and women (in inches) in a certain population can be regarded as normal random variables with the respective distributions $N(69,9)$ and $N(66,7)$. A man and a woman are drawn at random from the population. Find the probability that the woman is taller than the man.

Solution. Let X and Y be the random variables giving the heights of men and women respectively. Then X is $N(69,9)$ and Y is $N(66,7)$. Now notice that the random variable $-Y$ is $N(-66,7)$: this is because the p.d.f. of $-Y$ is just the reflection of the p.d.f. of Y in the y-axis, so that the mean is -66 and the variance (as a measurement of the spread of the observations) is unaltered at 7. So, applying the above result to X and $-Y$, we conclude that $Z = X - Y$ is a normal random variable with the distribution $N(3,16)$. The question asks us for $P(Y > X)$, i.e. for $P(Z < 0)$. So the answer $= P(Z < 0) = P\left(\dfrac{Z-3}{4} < \dfrac{-3}{4}\right)$

$$= P\left(Q < -\dfrac{3}{4}\right), \text{ where } Q = \dfrac{Z-3}{4} \text{ is}$$
$$= \Phi(-0.75) \qquad\qquad N(0,1),$$
$$= 1 - \Phi(0.75) = 1 - 0.7734 = 0.2266.$$

EXAMPLES TO DO: Pages 88 - 89: Exs. 10 - 12.

EXAMPLES 10

[The $N(0,1)$ table required is given on page 118.]

1. Let X be an $N(0,1)$ random variable. Find

    (a) $P(0 < X < 1)$,   (b) $P(0.5 < X < 1.3)$,
    (c) $P(-0.7 < X < 0.7)$, (d) $P(X > 1.5)$,
    (e) $P(X \leq -2)$,   (f) $P(X \geq 1.5 \mid X > 1)$.

2. The random variable X has the normal distribution $N(100, 64)$. Find

    (a) $P(X > 102)$,   (b) $P(90 < X < 110)$,
    (c) $P(X < 110 \mid X > 102)$.

If three observations of X are made, what is the probability that exactly two of them are greater than 102?

3. The highest June temperature (in °C) at a certain spot in any given year is a normal random variable X with the distribution $N(20, 16)$. Find

    (a) $P(18 < X < 21)$,   (b) $P(X > 19)$,
    (c) $P(X > 25 \mid X > 19)$.

4. The weight of a certain type of bird (in grams) has approximately the normal distribution $N(100, 256)$. Find the probability that

    (a) a bird chosen at random weighs more than 112 grams,
    (b) a bird chosen at random weighs between 92 and 104 grams.

If four birds are weighed, what is the probability that at least three weigh more than 104 grams?

If six birds are weighed one after another, what is the probability that exactly five successive birds weigh more than 104 grams?

5. In a national examination for schools it is found that English marks can be regarded as having an $N(50, 144)$ distribution while Mathematics marks can be regarded as having an $N(55, 400)$ distribution. By scaling both marks to $N(0,1)$, decide which is better 63 in English or 76 in Mathematics.

6. The life length of a certain type of light bulb has the distribution $N(200, 400)$, where the life length is measured in hours. Find

    (a) $P(X \geq 216)$,   (b) $P(180 \leq X \leq 220)$,

(c) $P(X > 224 \mid X > 216)$.

A second type of light bulb has life length (in hours) with the normal distribution $N(210,144)$. Which type of bulb is more likely to survive a period of 230 hours?

7. The time taken (in days) by a shipyard A to build a standard cargo ship can be regarded as a random variable with the normal distribution $N(300,625)$. Find the probability that the time taken for a particular ship is

(a) more than 330 days,
(b) between 285 and 320 days,
(c) less than 300 days given it is over 280 days.

The time taken (in days) by a second shipyard B to build the same type of ship is a random variable which is uniformly distributed on the interval [255, 355]. State (giving reasons) which shipyard is more likely to complete this type of ship within 275 days.

8. Light bulbs have life length X hours, where X can be regarded as an $N(1200,2500)$ random variable. Find the probability that a bulb lasts for between 1180 and 1240 hours. If three bulbs are inserted in a room (all burning simultaneously) what is the probability that a light will still be burning in the room after 1280 hours?

9. Two builders A and B are available to build a standard type of house. The times taken to complete such a house (in days) can be taken as normal random variables with the distributions $N(130,100)$ for A and $N(140,324)$ for B. Find which builder is more likely to have completed his house in a period of

(a) 145 days, (b) 115 days, (c) 110 days.

Both builders start at the same time, working independently, each building a house. Find the probability that at least one of the houses will be completed within 122 days.

10. Two light bulbs, one of each type in Ex. 6 above, are set burning together. Find the probability that the bulb of the second type will last longer.

11. In the last part of Ex. 9 above, find the probability that builder B finishes his house first.

12. A river buoy flashes continuously. When it is serviced it is fitted with five new bulbs arranged so that one bulb provides the light until it burns out, the second bulb then being tripped into service until it also fails, the third bulb then being tripped and so on.

Once all five bulbs have failed the buoy becomes unlit (and must be serviced immediately). The life length (in days) of each bulb can be regarded as a normal random variable with mean 50 and standard deviation 20. Find the probability that the buoy will become unlit within
        (a) 180 days, (b) 200 days.

XI. TWO DIMENSIONAL RANDOM VARIABLES

85. In Chapter 5 we discussed one dimensional random variables - there being two basic types, discrete and continuous. In this chapter we discuss briefly the idea of two dimensional random variables - we restrict our attention to the continuous case. The discrete case can be dealt with by analogy.

86. THE BACKGROUND

Think of a submarine captain hunting for a ship on an ocean. He may know that on a grid of coordinates the ship was at the origin point (0,0) two hours ago, and he may be wondering about its current position (X,Y) Obviously there is a distribution of probability (or "probability sand" - see §6) across the map, the entire probability lying within two hours steaming distance of the origin.

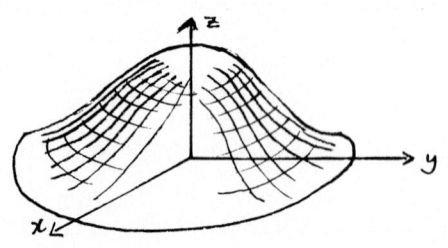

Effectively this means that we can think of a probability density function $f(x,y)$, i.e. a function of two variables with the surface $z = f(x,y)$ giving the distribution of probability, as shown in the diagram. The probability of the ship lying in a particular region A then corresponds to the amount of "probability sand" (i.e. the volume under the surface) contained above the plane $z = 0$, below the surface $z = f(x,y)$ and above the two dimensional region A. The definition in §87 makes these ideas precise.

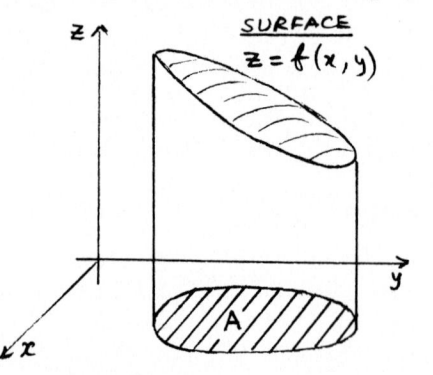

87. JOINT PROBABILITY DENSITY FUNCTION

<u>Definition</u>. Let (X,Y) be a continuous two dimensional random variable taking values in R^2 (i.e. in real two

dimensional space or the xy-plane). Then the joint probability density function f for (X,Y) is a function of two variables satisfying the following conditions:

(a) $f(x,y) \geq 0$ for all $(x,y) \in R^2$;

(b) $\iint_{R^2} f(x,y) \, dxdy = 1$;

(c) For every region A in R^2,

$$P\left[(X,Y) \in A\right] = \iint_A f(x,y) \, dxdy \, .$$

Notice: (i) Much of the theory here is similar to the theory for the one dimensional case in §41, except that double integration replaces single integration.

(ii) In practice f(x,y) may take the value 0 in parts of the xy-plane. For example, barring the supernatural, the position (X,Y) of the ship in §86 could not be more than two hours steaming distance from the origin (say r miles). So f(x,y) would certainly be zero for $x^2 + y^2 > r^2$.

This means that the region of integration in (b) can in practice be reduced to the region on which $f(x,y) > 0$.

(iii) The amount of "probability sand" associated with a region A of the xy-plane is (as explained in §86) represented by the volume under the surface z = f(x,y), above the xy-plane, and inside a cylinder with A as its base and axis parallel to the z-axis. (In plain language this means that you can think of the probability as the volume cut out by a scone cutter with the boundary of A as its rim.)

(iv) The condition (b) above corresponds to the fact that the total probability (i.e. total volume contained between the surface z = f(x,y) and the xy-plane is 1 unit.

88. EXAMPLE. The two dimensional random variable (X,Y) has the joint p.d.f. given by

$$f(x,y) = \begin{cases} \frac{1}{20}(x + 4y) & (0 \leq x \leq 2 \text{ and } 0 \leq y \leq 2), \\ 0 & \text{otherwise.} \end{cases}$$

Find (a) $P(X < 1)$,   (b) $P(X + Y \geq 1)$.

Solution. Clearly (X,Y) takes only values in the square with vertices at (0,0), (2,0), (2,2) and (0,2).

So we can restrict our attention to this square. The events required are shaded in the diagrams below. We find the required probabilities by doing double integration of the joint p.d.f. over these shaded regions.

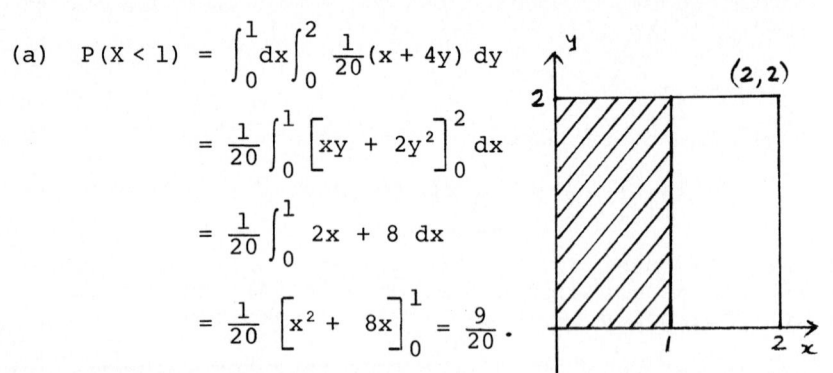

(a) $P(X < 1) = \int_0^1 dx \int_0^2 \frac{1}{20}(x + 4y)\, dy$

$= \frac{1}{20} \int_0^1 \left[xy + 2y^2 \right]_0^2 dx$

$= \frac{1}{20} \int_0^1 2x + 8\, dx$

$= \frac{1}{20} \left[x^2 + 8x \right]_0^1 = \frac{9}{20}.$

(b) Here it is easier to find $P(X + Y < 1)$, i.e. the complementary event (the unshaded area) and subtract this value from 1.

$P(X + Y < 1) = \int_0^1 dx \int_0^{1-x} \frac{1}{20}(x + 4y)\, dy$

$= \frac{1}{20} \int_0^1 \left[xy + 2y^2 \right]_0^{1-x} dx$

$= \frac{1}{20} \int_0^1 x(1-x) + 2(1-x)^2\, dx$

$= \frac{1}{20} \int_0^1 x^2 - 3x + 2\, dx$

$= \frac{1}{20} \left[\frac{x^3}{3} - \frac{3x^2}{2} + 2x \right]_0^1 = \frac{1}{24}.$

So we conclude that $P(X + Y \geq 1) = 1 - \frac{1}{24} = \frac{23}{24}.$

<u>Notice</u>: The amount of probability associated with a line segment in the sample space for a continuous two dimensional random variable is zero. For example, in the above example, $P(X = 1)$ is zero. Consequently, from part (a), we can conclude that

$$P(X \leq 1) = P(X < 1) = \frac{9}{20}.$$

Compare Remark (iii) in §42 for the one dimensional case.

<u>EXAMPLES TO DO</u>: Page 96: Exs. 1, 2, 3.

89. INDEPENDENT RANDOM VARIABLES

Obviously we would like X and Y to be called independent if and only if the value of X has no influence on the value of Y and vice versa. It is beyond the scope of this book to give a formal definition of this concept: this would need a discussion of what are known as marginal distributions. Look at more advanced books if you are interested in this.

We will however state the following result, which will be of use in the examples in §90 and §91.

RESULT. Suppose that X and Y are independent one dimensional random variables with respective probability density functions f and g. Then the joint probability density function of the two dimensional random variable (X,Y) is given by h where

$$h(x,y) = f(x)g(y) \quad \text{for} \quad (x,y) \in R^2.$$

Notice: In other words, if X and Y are independent then the joint p.d.f. is the product of the individual probability density functions for X and Y.

90. EXAMPLE.
The yearly outputs of two mines A and B can be regarded as independent random variables X and Y with respective probability density functions

$$f(x) = \begin{cases} e^{-x} & (x > 0), \\ 0 & \text{otherwise,} \end{cases} \quad g(y) = \begin{cases} 4e^{-4y} & (y > 0), \\ 0 & \text{otherwise.} \end{cases}$$

Find the probability that in a given year

(a) the output of mine A exceeds that of mine B,

(b) the sum of the outputs of the two mines is greater than 1 unit.

Solution. The joint p.d.f. is (by §89)

$$h(x,y) = \begin{cases} 4e^{-(x+4y)} & (x > 0, y > 0), \\ 0 & \text{otherwise.} \end{cases}$$

(a) Here the required event is given by those points (x,y) in the plane for which $x > y$. This region is shaded in the diagram overleaf. Doing double integration over this region gives

$$P(X > Y) = \int_0^\infty dx \int_0^x 4e^{-x}e^{-4y}\, dy$$

$$= \int_0^\infty e^{-x} \cdot \left[-e^{-4y}\right]_0^x dx$$

$$= \int_0^\infty e^{-x}\left(1 - e^{-4x}\right) dx$$

$$= \int_0^\infty e^{-x} - e^{-5x}\, dx$$

$$= \left[-e^{-x} + \tfrac{1}{5}e^{-5x}\right]_0^\infty$$

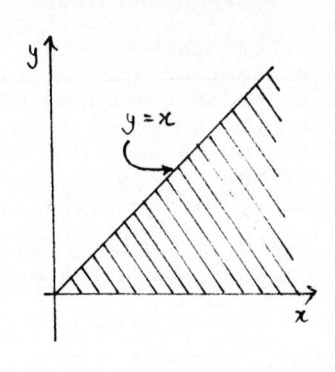

$$= 1 - \tfrac{1}{5} = \tfrac{4}{5}.$$

(b) The event is given by the set of points for which $x + y > 1$ (i.e. the shaded region in the diagram). It is easier from the point of view of integration to find $P(X + Y \leqq 1)$ and subtract this from 1.

$$P(X + Y \leqq 1) = \int_0^1 dx \int_0^{1-x} 4e^{-(x+4y)}\, dy$$

$$= \int_0^1 e^{-x} \cdot \left[-e^{-4y}\right]_0^{1-x} dx$$

$$= \int_0^1 e^{-x} - e^{3x-4}\, dx$$

$$= \left[-e^{-x} - \tfrac{1}{3}e^{3x-4}\right]_0^1$$

$$= 1 - \tfrac{4}{3}e^{-1} + \tfrac{1}{3}e^{-4}.$$

So $P(X + Y > 1) = 1 - P(X + Y \leqq 1)$ (since total probability equals 1)

$$= \tfrac{4}{3}e^{-1} - \tfrac{1}{3}e^{-4}.$$

EXAMPLES TO DO: Pages 96 - 97:   Exs. 4, 5.

91. SUPPLY AND DEMAND PROBLEMS

We illustrate with an example.

EXAMPLE. The daily supply of milk to a town is a continuous random variable X with p.d.f. given by

$$f(x) = \begin{cases} \tfrac{1}{2}x^2 e^{-x} & (x \geq 0), \\ 0 & \text{otherwise.} \end{cases}$$

The daily demand for milk is a continuous random variable Y, independent of X, and with p.d.f. given by

$$g(y) = \begin{cases} e^{-y} & (y \geq 0), \\ 0 & \text{otherwise.} \end{cases}$$

Find the probability that on a given day demand exceeds supply.

Solution. The joint p.d.f. is given by h where

$$h(x,y) = \begin{cases} \tfrac{1}{2}x^2 e^{-(x+y)} & (x \geq 0 \text{ and } y \geq 0), \\ 0 & \text{otherwise.} \end{cases}$$

The event of interest is given by y > x (shaded area).

$$P(Y > X) = \int_0^\infty dx \int_x^\infty \tfrac{1}{2}x^2 e^{-(x+y)} \, dy$$

$$= \int_0^\infty \tfrac{1}{2}x^2 e^{-x} \left[-e^{-y}\right]_x^\infty dx$$

$$= \int_0^\infty \tfrac{1}{2}x^2 e^{-2x} \, dx$$

$$= \frac{1}{16} \int_0^\infty u^2 e^{-u} \, du \quad \text{(on setting } u = 2x\text{)}$$

$$= \frac{1}{16}\Gamma(3) \quad \text{(See the Gamma function in §100)}$$

$$= \frac{2}{16} = \frac{1}{8}.$$

Notice: Here we could avoid the Gamma function and use integration by parts to evaluate the final integral. However the Gamma function is a great help with such integrals in general.

EXAMPLES TO DO: Pages 96 - 97: Exs. 6, 7, 8.

EXAMPLES 11

1. A two dimensional random variable (X,Y) is uniformly distributed over the square with vertices at $(0,0)$, $(4,0)$, $(4,4)$, $(0,4)$. Find
    (a) the joint p.d.f. of (X,Y),
    (b) $P(X > 1)$,
    (c) $P(X + Y > 1)$,
    (d) $P(X < 3 \mid X + Y > 1)$.

2. A two dimensional random variable (X,Y) has the joint probability density function given by

$$f(x,y) = \begin{cases} a(3x^2 + 2xy) & (0 \leq x \leq 4 \text{ and } 0 \leq y \leq 1), \\ 0 & \text{otherwise,} \end{cases}$$

where a is a positive constant. Find
(a) the constant a, (b) $P(X < 2)$, (c) $P(X + Y < 1)$.

3. A two dimensional random variable (X,Y) has the joint p.d.f. given by

$$f(x,y) = \begin{cases} \frac{3}{16} x^2 y & (0 \leq x \leq 2 \text{ and } 0 \leq y \leq 2), \\ 0 & \text{otherwise.} \end{cases}$$

Find (a) $P(X \leq 1)$, (b) $P(Y \leq 1)$,
    (c) $P(\max(X,Y) \leq 1)$, (d) $P(\min(X,Y) \geq 1)$,
    (e) $P(\max(X,Y) > 1)$, (f) $P(\min(X,Y) < 1)$.

4. The weekly outputs of two copper mines H and K can be regarded as independent random variables X and Y with respective probability density functions

$$f(x) = \begin{cases} 9xe^{-3x} & (x > 0), \\ 0 & \text{otherwise,} \end{cases} \qquad g(y) = \begin{cases} 4e^{-4y} & (y > 0), \\ 0 & \text{otherwise.} \end{cases}$$

Find the probability that in a given week the output of mine K exceeds that of mine H. Find also the probability that in a given week the sum of the outputs of the two mines will be less than 1 unit.

5. Independent continuous random variables X and Y have respective probability density functions given by

$$f(x) = \begin{cases} 2xe^{-x^2} & (x > 0), \\ 0 & \text{otherwise,} \end{cases} \qquad g(y) = \begin{cases} 2y & (0 < y < 1), \\ 0 & \text{otherwise.} \end{cases}$$

Find $P(Y \geq X)$.

6. Two opencast mines A and B operate independently and their monthly productions (in suitable units) are random variables X and Y with respective probability density functions

$$f(x) = \begin{cases} xe^{-x} & (x \geq 0), \\ 0 & \text{otherwise,} \end{cases} \qquad g(y) = \begin{cases} 4ye^{-2y} & (y \geq 0), \\ 0 & \text{otherwise.} \end{cases}$$

Find (a) $P(Y \leq 2)$,
(b) the probability that in a given month the production of mine B exceeds that of mine A.

Denoting the answer to (b) by q, find the probability in terms of q that the production of mine B exceeds the production of mine A in at least one month in the year.

7. The daily supply of water to a power station is a continuous random variable X with p.d.f. given by

$$f(x) = \begin{cases} 2xe^{-x^2} & (x \geq 0), \\ 0 & (x < 0). \end{cases}$$

Using the Gamma function, evaluate $E(X)$ and $V(X)$.

Daily demand for water in the power station is a random variable Y (independent of X) and with p.d.f.

$$g(y) = \begin{cases} 8ye^{-4y^2} & (y \geq 0), \\ 0 & (y < 0). \end{cases}$$

Find the probability that on a given day demand exceeds supply.

8. At a fruit market, the daily supply X and the daily demand Y for strawberries can be regarded as independent random variables with respective probability density functions given by

$$f(x) = \begin{cases} 4x^2 e^{-2x} & (x \geq 0), \\ 0 & \text{otherwise,} \end{cases} \qquad g(y) = \begin{cases} 3e^{-3y} & (y \geq 0), \\ 0 & \text{otherwise.} \end{cases}$$

Find the probability that on a given day demand exceeds supply.

XII. CHEBYSHEV'S INEQUALITY AND THE CENTRAL LIMIT THEOREM

92. CHEBYSHEV'S INEQUALITY

In §81 we saw that for a <u>normal distribution</u> roughly 95% of the probability <u>lies within 2 standard deviations of the mean</u>, i.e. for an $N(\mu, \sigma^2)$ distribution roughly 95% of the probability lies in the interval $(\mu - 2\sigma, \mu + 2\sigma)$.

Knowing how much probability lies within so many standard deviations of the mean is a matter of considerable interest. A standard general result in this area is Chebyshev's Inequality: it applies to <u>all</u> distributions with finite mean and variance. We state it without proof as the following result.

<u>RESULT</u>. (<u>Chebyshev's Inequality</u>) Let X be any random variable with $E(X) = \mu$ and $V(X) = \sigma^2$, (so that the standard deviation is σ). Then

(i) for every positive real number k,

$$P\left(|X - \mu| \geq k\sigma\right) \leq \frac{1}{k^2},$$

or equivalently, on considering the complementary event,

(ii) for every positive real number k,

$$P\left(|X - \mu| < k\sigma\right) \geq 1 - \frac{1}{k^2}.$$

<u>Proof</u>. See more advanced books.

<u>Notice</u>: (i) The number k is the number of standard deviations from the mean. You can set k to any positive value.

(ii) If we take $k = 2$ in the second form of the result, we then conclude that

$$P\left(|X - \mu| < 2\sigma\right) \geq \tfrac{3}{4},$$

i.e. the probability that X lies in the interval $(\mu - 2\sigma, \mu + 2\sigma)$ is at least $\tfrac{3}{4}$. So, in the diagram, we are guaranteed that the amount of probability in the shaded central region is at least $\tfrac{3}{4}$. And remember this applies <u>no matter what the distribution</u> (provided that it has finite mean and variance).

Correspondingly the sum of the amounts of probability in the two unshaded tails (i.e. less than or equal to $\mu - 2\sigma$ or greater than or equal to $\mu + 2\sigma$) is at most $\tfrac{1}{4}$.

$y = f(x)$ is the p.d.f.

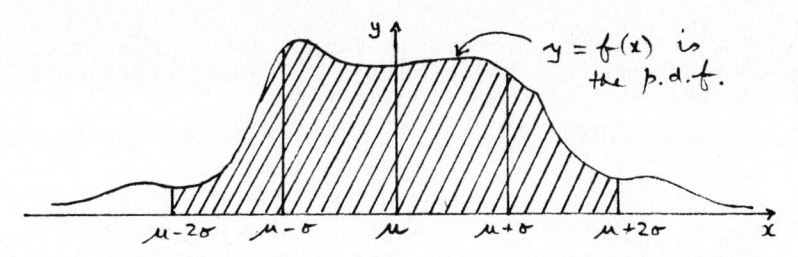

Similarly taking k = 3 shows that the probability of lying within 3 standard deviations of the mean is at least $8/9$, and so on. Taking k = 1 is fruitless.

(iii) Be clear that the result applies to every random variable with finite mean and variance and that there do exist random variables for which the inequality is sharp, i.e. there do exist distributions for which

$$P\left(|X - \mu| \geq k\sigma\right) = \frac{1}{k^2}.$$

See Ex.7 at the end of this chapter for more on this.

93. EXAMPLE. Two types of light bulb A and B have life lengths (X hours) which are random variables with means and standard deviations as given in the following table:

type	mean μ	standard deviation σ
A	500	5
B	500	20

Use Chebyshev's Inequality to make a statement about P(460 < X < 540) in each case.

Solution. We use the notation μ, σ, k as in the result in §92.

For A, take μ = 500, σ = 5, k = 8. So by part (ii) of the result in §92 we have (for A)

$$P(460 < X < 540) \geq 1 - \frac{1}{8^2} = \frac{63}{64}.$$

For B, taking μ = 500, σ = 20, k = 2 gives

$$P(460 < X < 540) \geq 1 - \frac{1}{2^2} = \frac{3}{4}.$$

94. If you toss a coin 1600 times and obtain only 703 heads, do you have reason to feel cheated? The following example, which uses Chebyshev's Inequality, sheds some light on this question.

EXAMPLE. An unbiased coin is tossed 1600 times, the resulting number of heads being X. Use Chebyshev's Inequality to find lower bounds for

    (a)   P(760 < X < 840),   (b)   P(720 < X < 880),

    (c)   P(700 < X < 900).

Solution. The random variable X is binomial with $n = 1600$ and $p = \frac{1}{2}$. From §63, we know that

$$E(X) = np = 800, \quad V(X) = np(1-p) = 400.$$

So in Chebyshev's Inequality (§92) take $\mu = 500$ and $\sigma = 20$. For the three cases required we take the value of k to be 2, 4, 5 respectively. This gives

(a) $P(760 < X < 840) = P(|X - 800| < 2\sigma) \geq \frac{3}{4}$;

(b) $P(720 < X < 880) = P(|X - 800| < 4\sigma) \geq \frac{15}{16}$;

(c) $P(700 < X < 900) = P(|X - 800| < 5\sigma) \geq \frac{24}{25}$.

Notice: (i) In theory you can calculate the exact answers using the formula for the binomial distribution from §37. For example the exact value for (a) is

$$\sum_{k=761}^{839} \binom{1600}{k} (\tfrac{1}{2})^k (\tfrac{1}{2})^{1600-k} = (\tfrac{1}{2})^{1600} \sum_{k=761}^{839} \binom{1600}{k}.$$

The arithmetic is colossal.

    (ii) The estimates given by Chebyshev in this problem are actually quite crude, though they have the advantage of being easy to calculate. This crudeness is the price we pay for the broad spectrum of distributions to which Chebyshev applies. In this problem we actually know that the distribution is binomial, but using Chebyshev's Inequality does not really capitalise on this.

    It is possible to obtain more accurate estimates in this problem using the normal approximation to the binomial: see §98 - 99 and Ex. 1 at the end of the chapter.

EXAMPLES TO DO: Page 108: Ex. 7.

95. THE BACKGROUND TO THE CENTRAL LIMIT THEOREM

Suppose we throw a dice ten times, the scores being $X_1, X_2, \ldots, X_{10}$. Each individual X_i then lies in the range $1 \leq X_i \leq 6$ ao that the total score $(X_1 + X_2 + \ldots + X_{10})$ lies in the range given by $10 \leq X_1 + X_2 + \ldots + X_{10} \leq 60$.

Actually each $E(X_i) = 3.5$ so that by §56 we can say that $E(X_1 + X_2 + \ldots + X_{10}) = 35$. It is fairly clear however that the sum $X_1 + X_2 + \ldots + X_{10}$ is more likely to be near 35 than it is to be near 10 or near 60. For evidence of this compare two cases - the case when the sum is exactly 10 and the case where the sum is exactly 35. On one hand there is only <u>one</u> way that the sum $X_1 + X_2 + \ldots + X_{10}$ can be equal to 10, namely $X_i = 1$ for every i. On the other hand there are lots of ways that the sum $X_1 + X_2 + \ldots + X_{10}$ can be equal to 35 : for a start we can have all permutations of the sequence (3, 3, 3, 3, 3, 4, 4, 4, 4, 4) [252 permutations in all] and of course there are lots of other possibilities using the outcomes 1, 2, 5 and 6 as well.

So, <u>even though</u> for each individual throw of the dice, the probability of each outcome 1, 2, 3, 4, 5, 6 is equal to $\frac{1}{6}$ (i.e. the distribution is <u>flat</u>), the distribution of the sum $X_1 + X_2 + \ldots + X_{10}$ is <u>not flat</u>: in fact it has a central peak at 35 and falls away steadily on the left towards 10 and on the right towards 60. The truth is (and this is the plain language message of the Central Limit Theorem) that the distribution of $X_1 + X_2 + \ldots + X_{10}$ round the mean value of 35 is <u>approximately a normal distribution.</u>

96. THE CENTRAL LIMIT THEOREM

We state the Central Limit Theorem as the result which follows. Be warned that the quality of the approximation to the normal distribution in the result depends in general on the number of individual random variables X_i (i.e. it depends on the number n). As a rule the quality of the approximation improves as the value of n increases. Some comments about how large n should be are made in the notes which follow the result.

RESULT. (Central Limit Theorem) Let $X_1, X_2, \ldots, X_n$ be n independent identically distributed random variables each with mean μ and standard deviation σ. Then, provided that n is sufficently large [see (ii) below], we have that

(i) the random variable Y given by

$$Y = X_1 + X_2 + \ldots + X_n$$

has approximately the normal distribution $N(n\mu, n\sigma^2)$,

<u>and</u> equivalently,

(ii) the random variable Z given by

$$Z = (X_1 + X_2 + \ldots + X_n - n\mu)/(\sigma\sqrt{n})$$

is approximately $N(0,1)$.

<u>Proof</u>. See more advanced books.

<u>Notice</u>: (i) No matter what the distribution of the original random variables X_i (provided that the mean and standard deviation are finite) the above result still guarantees that the distribution of the sum is approximately normal.

(ii) More precise statements of the result say that the distribution of the random variable $X_1 + X_2 + \ldots + X_n$ tends to the normal distribution as $n \to \infty$. In practice however we will want to use the result for finite values of n. Just how large n should be to be sure of a close approximation to normal behaviour is difficult to say precisely: it depends also on the shape of the original distributions. Certainly for $n > 30$ we have reasonable hope of good approximation, but these matters are rather beyond the scope of this book. We can also apply the result for $n \leq 30$ (e.g. $n = 10$ in §95) <u>provided that</u> we treat the results with some degree of caution, the smaller the value of n the more the caution.

(iii) The frequent occurrence of the normal distribution in real situations can be attributed with some justification to the Central Limit Theorem. For example, the daily consumption of water in a town may follow a roughly normal distribution from day to day: the total consumption can be regarded as a sum of many independent random variables (of unknown distribution) (i.e. the consumptions by individual people). The Central Limit Theorem suggests that the total consumption will be roughly normal. Likewise for the total error in a quantity calculated as the sum of individual quantities.

97. **EXAMPLE.** A dice is thrown 100 times. Use the Central Limit Theorem to estimate the probability that the total score will lie between 320 and 380.

Solution. For each individual throw of the dice, the result is X, where

$$E(X) = \frac{1}{6}(1+2+3+4+5+6) = 3.5,$$

and $E(X^2) = \frac{1}{6}(1+4+9+16+25+36) = \frac{91}{6} = 15.167$.

So this gives $V(X) = 15.167 - (3.5)^2 = 2.917$, which gives the standard deviation $\sigma(X) = 1.708$.

Letting $Y = X_1 + X_2 + \ldots + X_{100}$, we apply the Central Limit theorem with $\mu = 3.5$, $\sigma = 1.708$ and $n = 100$. So $Z = \frac{Y - 350}{17.08}$ is approximately $N(0,1)$. The question asks us to find $P(320 \leq Y \leq 380)$ but because we are approximating a <u>discrete</u> random variable (i.e. Y) by a <u>continuous</u> one (i.e. the normal distribution) we make the so called <u>correction for continuity</u> (explained below) and find $P(319.5 \leq Y \leq 380.5)$ instead. So

$$P(320 \leq Y \leq 380) = P(319.5 \leq Y \leq 380.5)$$

$$= P\left(\frac{319.5 - 350}{17.08} \leq Z \leq \frac{380.5 - 350}{17.08}\right),$$

(where Z is approximately $N(0,1)$),

$$= P(-1.786 \leq Z \leq 1.786)$$

$$= \Phi(1.786) - \Phi(-1.786) \quad \text{(approximately)}$$

$$= 0.926 \quad \text{(from } N(0,1) \text{ tables)}.$$

Notice: (The <u>correction for continuity</u>) In the above example, the random variable Y given by

$$Y = X_1 + X_2 + \ldots + X_{100}$$

is <u>discrete</u> and can take all <u>integer</u> values from 100 to 600. The Central Limit Theorem tells us to approximate Y by the normal distribution (a <u>continuous</u> distribution) with mean 350 and standard deviation 17.08. So we are approximating a <u>discrete</u> distribution with a <u>continuous</u> one, and while the random variable Y cannot take values like 350.27 the approximating normal random variable can. In the histogram on the left P(Y = 349) is represented by the area of the shaded strip centred at the point 349. (Notice that the length of the base of each strip rectangle has been taken as 1 unit, e.g. the rectangle corresponding to P(Y = 349) straddles the interval from 348.5 to 349.5. So the area of each strip is numerically equal to its height.)

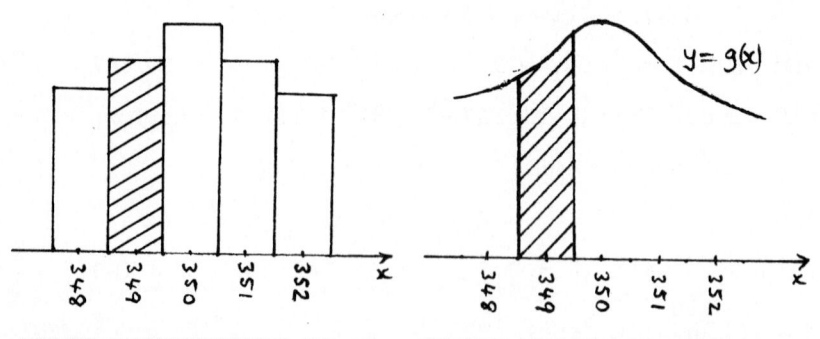

EXACT DISTRIBUTION OF Y   NORMAL APPROXIMATION TO Y

By analogy, in the approximation using the normal distribution (say with p.d.f. called g) we estimate P(Y = 349) as the strip of area centred at 349 and with base of length 1 unit (diagram on the right), i.e. as

$$P(Y = 349) = \int_{348.5}^{349.5} g(y)\, dy.$$

This <u>correction for continuity</u> is required because the normal distribution, which is providing the approximation, is continuous and attempting to find P(Y = 349) using

$$\int_{349}^{349} g(y)\, dy$$

gives the answer zero. (See Remark (iii) in §42.) In plain language the correction for continuity means "add ½ at each end of the interval to include probability which rightfully belongs to the endpoints".

This point also arises in connection with the normal approximation to the binomial in §98 and §99.

EXAMPLES TO DO: Page 107: Exs. 3, 4.

98.   THE NORMAL APPROXIMATION TO THE BINOMIAL

This is a special case of the Central Limit Theorem.   We state it as a result.

RESULT. (Normal approximation to the binomial) Let X be a binomial random variable based on n repetitions and with constant probability p of success. Then for suitably large values of n (see (i) below), X has approximately the normal distribution with mean np and standard deviation $\sqrt{(np(1-p))}$.

Proof. We can regard X as $X = X_1 + X_2 + \ldots + X_n$, where each X_i is given by

$$X_i = \begin{cases} 1 & \text{if there is success on the ith repetition,} \\ 0 & \text{if there is failure on the ith repetition.} \end{cases}$$

As such X is the sum of n independent identically distributed random variables - the classic setting for the Central Limit Theorem. Also we know (from §63) that $E(X_i) = p$ and $V(X_i) = p(1-p)$. So the Central Limit Theorem tells us that, for large values of n, the distribution of $X = X_1 + X_2 + \ldots + X_n$ is approximately normal with mean np and variance $np(1-p)$.

Notice: (i) The result talks of the approximation being valid for suitably large values of n. In fact, the value of p also affects the quality of the approximation.

Generally speaking the approximation is better if the value of p is close to $\frac{1}{2}$ (rather than close to 0 or close to 1). If the value of p is close to $\frac{1}{2}$ then the value of n can be reasonably be as low as 20, but if p is close to 0 or close to 1 then n will have to be much larger. A fairly standard guide is that the normal approximation to the binomial can reasonably used if both np and $n(1-p)$ are both greater than 5.

(ii) The Poisson approximation to the binomial in §71 applies in circumstances rather different to those described in (i) for the normal approximation. For the Poisson approximation we need p to be close to 0 (not close to $\frac{1}{2}$) as well as requiring n to be large. See §71 on this.

(iii) (Correction for continuity) In using the normal approximation to the binomial we are using a continuous distribution to approximate a discrete one and the correction for continuity as described in §97 is required in doing examples. The example in §99 is a case in point.

99. **EXAMPLE.** An unbiased coin is tossed 100 times. Let X be the number of heads obtained. Using the normal approximation to the binomial, estimate

(a) $P(X = 48)$,   (b) $P(52 \leq X \leq 55)$.

Solution. Here X is binomial with $n = 100$ and $p = \frac{1}{2}$. So $np = 50$ and $n(1-p) = 50$ and we can therefore use the normal approximation (by (i) in §98) which will have mean np and standard deviation $\sqrt{(np(1-p))}$, i.e. mean 50 and standard deviation 5. So we conclude that $Y = \frac{X - 50}{5}$ is approximately $N(0,1)$. So we make the following estimates (using the correction for continuity as explained in §97).

(a) $P(X = 48) = P(47.5 \leq X \leq 48.5)$

$= P\left(\frac{47.5 - 50}{5} \leq \frac{X - 50}{5} \leq \frac{48.5 - 50}{5}\right)$

$= P(-0.5 \leq Y \leq -0.3)$,
     (where Y is approximately $N(0,1)$),

$= \Phi(-0.3) - \Phi(-0.5)$ (approximately)

$= 0.0736$ (from $N(0,1)$ tables).

(b) $P(52 \leq X \leq 55) = P(51.5 \leq X \leq 55.5)$

$= P\left(\frac{51.5 - 50}{5} \leq \frac{X - 50}{5} \leq \frac{55.5 - 50}{5}\right)$

$= P(0.3 \leq Y \leq 1.1)$,
     (where Y is approximately $N(0,1)$),

$= \Phi(1.1) - \Phi(0.3)$ (approximately)

$= 0.2464$ (from $N(0,1)$ tables).

EXAMPLES TO DO: Page 107: Exs. 1, 2, 5, 6.

EXAMPLES 12

[The $N(0,1)$ table required is given on page 118.]

1. A coin is tossed 1600 times. Let X be the number of heads obtained. Use the normal approximation to the binomial to estimate

(a) $P(760 < X < 840)$, (b) $P(720 < X < 880)$,
(c) $P(700 < X < 900)$.

Compare these estimates with the lower bounds obtained with Chebyshev's Inequality in §94.

2. A dice is thrown 180 times. Let X be the number of sixes obtained. Using the normal approximation to the binomial, find approximate values for

(a) $P(X \geq 45)$, (b) $P(25 \leq X \leq 35)$.

Can Chebyshev's Inequality contribute here and if so what does it offer? Also, write a computer program to find the exact value of $P(25 \leq X \leq 35)$. Compare.

3. A sideshow at a fairground consists of a gently sloping ramp with an array of 24 holes at the foot. The holes are marked with the numbers 1, 2, 3, 4, 5 and 6, there being four holes with each number and their arrangement is random. Contestants roll six balls down the ramp to fall into the holes, a person's total score being the sum of the corresponding numbers. Prizes are given for the following total scores: 7, 8, 10, 11, 12, 17, 25, 28, 32, 33, 35. Make some qualitative comments on this game in the light of the Central Limit Theorem.

4. (This example concerns the _dartboard_. For a tutorial on the layout of a dartboard look at Ex. 2 on page 64.) In a well known game two partners each throw three darts at a dartboard, their aim being to score a total of 101 or more between them (i.e. with a total of six darts). What (if anything) can the Central Limit Theorem say about this? Discuss also the cases when (a) your partner is an automaton and you are not, (b) you are both automata with the same strategy.

5. From a normal pack of 52 cards a card is drawn at random _with_ replacement 48 times. Let X be the number of hearts. Use the normal approximation to the binomial to estimate

(a) $P(X = 12)$, (b) $P(7 < X < 17)$, (c) $P(X \geq 17)$.

6. A plane has 95 seats. The airline decides to sell 100 tickets for a certain flight in the hope that some passengers will not turn up. From past experience they estimate that the probability that any given passenger will turn up is 0.9 and they assume that passengers behave independently. On this basis make an estimate of the probability that more than 95 passengers will turn up for the flight.

7. Let X be the discrete random variable taking the values -2, 0 and 2 with respective probabilities $\frac{1}{8}$, $\frac{3}{4}$ and $\frac{1}{8}$. Show that for X the inequality given by Chebyshev's Inequality is actually an equality for the case of $k = 2$ (in the notation given in §92). Construct a random variable to do the same when $k = 3$.

APPENDIX

100. A NOTE ON THE GAMMA FUNCTION

The value of the Gamma function Γ is defined, for every value of $k > 0$, by

$$\Gamma(k) = \int_0^\infty x^{k-1} e^{-x} \, dx.$$

So, for example,

$$\Gamma(1) = \int_0^\infty e^{-x} \, dx = \left[-e^{-x}\right]_0^\infty = 1,$$

$$\Gamma(3) = \int_0^\infty x^2 e^{-x} \, dx = \left[-(x^2 + 2x + 2)e^{-x}\right]_0^\infty = 2,$$

and so on. Working out such integrals using integration by parts or otherwise can be harrowing. The following properties of the Gamma function (which we state without proof) allow us to write down the values of many such integrals easily.

PROPERTIES OF THE GAMMA FUNCTION.

A. $\Gamma(1) = 1$, $\Gamma(2) = 1$, $\Gamma(3) = 2$, $\Gamma(4) = 6$ and in general $\Gamma(n) = (n-1)!$ for every positive integer n.

B. $\Gamma(\frac{1}{2}) = \sqrt{\pi}$.

C. $\Gamma(k) = (k-1)\,\Gamma(k-1)$ for all $k > 1$. Repeated application of this reduces the calculation of $\Gamma(k)$ to that of $\Gamma(k-p)$ where p is a positive integer and $0 < k-p \leq 1$. For example,

$$\Gamma(7/2) = \frac{5}{2} \cdot \frac{3}{2} \cdot \frac{1}{2}\,\Gamma(\tfrac{1}{2}) = \frac{15\sqrt{\pi}}{8}.$$

The following result gives a formula (in terms of the Gamma function) which allows you to write down the values of several integrals occurring in this book, by choosing suitable values of α, β and k.

RESULT. $\int_0^\infty x^{\alpha-1} e^{-kx^\beta}\,dx = \dfrac{\Gamma(\alpha/\beta)}{\beta\, k^{\alpha/\beta}}$, where α, β and k are all positive constants.

Proof. Setting $u = kx^\beta$ reduces the integral to a Gamma function.

As an example of the use of this last result, look at the calculation of the variance of the normal distribution in §83. There the integral

$$\int_0^\infty y^2 e^{-\frac{1}{2}y^2}\,dy$$

fits into this last result with $\alpha = 3$, $\beta = 2$ and $k = \frac{1}{2}$, and so has the value $\dfrac{\Gamma(3/2)}{2(\frac{1}{2})^{3/2}} = \dfrac{\frac{1}{2}\sqrt{\pi}}{2^{-\frac{1}{2}}} = \dfrac{\sqrt{\pi}}{\sqrt{2}}$.

Warning: The limits in such integrals must be 0 and ∞ for the formula in the above result to apply.

EXAMPLES TO DO: Pages 64 - 67: Exs. 3(e),(f),12.
Pages 96 - 97: Exs. 3 - 7.

HINTS AND ANSWERS TO THE EXAMPLES

[Note: References are to sections not to pages.]

EXAMPLES 1 (PAGES 7-8)

1. {5,6}, {1,3,5}, {3,4,5}, {8}, {7}.

2. {1,2,3,4,5,6, ... }.

3. {AB,BA,AC,CA,AD,DA,AE,EA,BC,CB,BD,DB,BE,EB, CD,DC,CE,EC,DE,ED}.

4. The set of all possible sequences of the letters Y and N of length 6, e.g. NNYNYN. Total number is 64.

5. S = {JK,JL,JM,JN,KL,KM,KN,LM,LN,MN,J,K,L,M,N,∅}.
(a) {JL,KL}, (b) {JN,J,KN,K,LN,L,MN,M,N,∅},
(c) {JK,KL,LM,MN}.

6. S = {(x,y): 100 ≤ x ≤ 300, 100 ≤ y ≤ 300}, i.e. S is the square in the xy-plane with vertices at the points (100,100), (300,100), (300,300), (100,300). The events are:

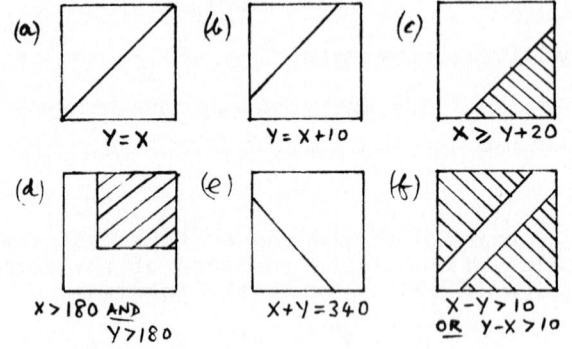

(a) Y = X
(b) Y = X+10
(c) X ≥ Y+20
(d) X > 180 AND Y > 180
(e) X+Y = 340
(f) X−Y > 10 OR Y−X > 10

7. Use the result in §7. Answers are 0.2, 0.1, 0.5.

8. Use the result in §7, noting P(A∪B) is at most 1 and is at least 0.8. So max = 0.5, min = 0.3.

9. Look first at the case n = 3 and generalise.

10. Number the eight disjoint pieces of a Venn diagram for the sets A, B, C and check that both sides consist of the same pieces. For the second part put C = B̄ in the first part.

11. Draw a Venn diagram and enter values. Answer is 0.20.

EXAMPLES 2 (PAGES 16-17)

1. 120.   2. 276.   3. 6840.   4. 1980.
5. Total = 5040;   (a) 720,   (b) 120,   (c) 240,   (d) 1440.
6. For the total number of words you really only need to choose the positions of the four letters B and this is the number of ways of choosing 4 out of 10, i.e. 210.   For the other parts, answers are 70 and 56.
7. 330; 150; 160.
8. (a) $\frac{21}{44}$,   (b) $\frac{21}{22}$.
9. (a) $\frac{3}{7}$,   (b) $\frac{19}{42}$,   (c) $\frac{13}{14}$.
10. (a) $\frac{8}{33}$,   (b) $\frac{2}{33}$.
11. Here you need to count cases with your bare hands. Answers: (a) $\frac{1}{66}$, (b) $\frac{1}{11}$, (c) $\frac{7}{66}$.
12. Similar to §15.   For n = 0, 1, 2, 3, 4 the answers reduce to 0.304, 0.439, 0.213, 0.041, 0.003.
13. Similar to §15.   On reduction with calculator or computer, answers are 0.0000164 and 0.00364.
14. (a) $\frac{5}{7}$,   (b) $\frac{1}{2}$.
15. $\frac{1}{21}$.   (Suppose one expert has chosen. Then ...)
16. (a) $\frac{1}{48}$,   (b) $\frac{5}{48}$.   Draw out all 144 cases in the sample space if you are in trouble here and count cases.

EXAMPLES 3 (PAGES 24-26)

1. $\frac{7}{13}$, $\frac{6}{13}$, $\frac{5}{13}$, $\frac{1}{3}$, $\frac{2}{5}$, $\frac{3}{7}$, $\frac{4}{7}$, $\frac{2}{13}$.
2. (a) $\frac{2}{11}$,   (b) $\frac{4}{55}$,   (c) $\frac{6}{55}$.
3. (a) $\frac{25}{102}$,   (b) $\frac{13}{51}$,   (c) $\frac{26}{51}$.
4. (a) $\frac{1}{216}$,   (b) $\frac{1}{72}$,   (c) $\frac{5}{9}$,   (d) $\frac{1}{36}$.
5. (a) $\frac{1}{3}$,   (b) $\frac{1}{15}$.   (See §21 for (b).)
6. Similar to §21.   Answer = $\frac{84}{715}$.

7. Similar to the the partition exercise in §22. Let Y, G, B be the events of the first ball being yellow, grey, blue. Let E = 2nd ball yellow. So $P(E) = P(E \cap Y) + P(E \cap G) + P(E \cap B)$, etc. Ans. = 6/11.

8. $\frac{24}{41}$. 9. $\frac{12}{37}$. 10. $\frac{8}{13}$.

11. Answers are $\frac{16}{17}$, $\frac{4}{5}$. For the second part you have to take both pieces of evidence into account, i.e. take E = First coin tosses 4 heads and second coin tosses two heads.

12. $\frac{53}{168}$, $\frac{28}{53}$, $\frac{343}{583}$. To do the last part, notice that once you know the ball drawn from Y is red, the first box can be in one of 2 states with probabilities $\frac{28}{53}$ and $\frac{25}{53}$. Then use the partition idea of §22.

13. $h^3/(h^3 + k^3)$, $h^3(1-k)^3/(h^3(1-k)^3 + k^3(1-h)^3)$. For the explicit values $\frac{27}{28}$, $\frac{729}{730}$. In the second part you have to use all the evidence so far available, as in the second part of Ex.11 above.

14. (a) $(5/6)^n$, (b) $5^{n-1}/6^n$.

15. (a) $\frac{1}{8}$, (b) $\frac{1}{5}$, (c) $\frac{1}{3}$. Look carefully at the difference between cases (b) and (c): writing out a list of pairs may be instructive.

EXAMPLES 4 (PAGES 33-34)

1. Apply the test for independence from §26.
(i) YES, (ii) YES, (iii) NO, (iv) YES, (v) YES.

2. (i) $\frac{13}{25}$, (ii) $\frac{2}{25}$, (iii) $\frac{3}{10}$, (iv) $\frac{19}{40}$, (v) $\frac{1}{4}$, (vi) $\frac{2}{5}$, (vii) $\frac{1}{2}$, (viii) $\frac{19}{49}$.

3. (a) $\frac{1}{16}$, (b) $\frac{7}{16}$ (Use §7), (c) $\frac{1}{4}$.

4. 0.20, $\frac{6}{11}$. 5. (a) 0.44, (b) $\frac{11}{13}$, (c) 0.6.

6. (a) $1 - (1-p^3)^2$, (b) $1 - (1-p^3)(1-p^2)(1-p)$, (c) $p + 2p^2 - 3p^3 + p^4$, (d) $[1-(1-p)^3][1-(1-p)^2]p$, (e) $[1-(1-p)^2][1-(1-p^2)^2]$.

9. $P(\overline{C} \cap \overline{D}) = P(\overline{C \cup D})$ by De Morgan (§1)
$= 1 - P(C \cup D)$. Now use result of §7 and independence.

EXAMPLES 6 (PAGES 41-43)

1. (a) $\frac{25}{216}$, (b) $\frac{57}{432}$. 2. (a) $\frac{15}{64}$, (b) $\frac{1}{64}$, (c) $\frac{1}{2}$.

3. (a) $\frac{448}{5^8}$, (b) $\frac{481}{5^8}$, (c) $1 - \frac{33}{5^8}$.

4. (a) $\frac{80}{243}$, (b) $\frac{15}{31}$, Last part = $\frac{448}{3^7}$.

5. (a) $\frac{8}{27}$, (b) $\frac{11}{27}$. Last part = $\frac{48 \cdot 65^2}{81^3}$ (Focus shifts as §39).

6. (a) $28p^2(1-p)^6$, (b) $1-(1-p)^7(1+7p)$. Last part: $10q^4(1-q)^2$.

7. (a) $10p^2(1-p)^3$, (b) $p^3(10-15p+6p^2)$. In the last part the focus changes from motors to units as in §39. Answer = $1-[1-(1-p)^4]^3$.

8. (a) $1 - \left(\frac{14}{15}\right)^6 = 0.339$, (b) $1-\left(\frac{13}{15}\right)^6 = 0.576$, (c) $1-\left(\frac{4}{5}\right)^6 = 0.738$, (d) $\frac{2}{5}$, (e) $\frac{4}{5}$.

9. $1 - \frac{7 \cdot 2^{12}}{3^{12}} = 0.946$. Last part = $\frac{34}{1443}$ (Note: Last part is not binomial.)

10. $1 - \left(\frac{21}{22}\right)^7 = 0.278$.

11. First part - Bayes - answer = $\frac{3}{19}$. Second part also Bayes but note proportions of boxes are again 20%, 70%, 10%. Evidence is 3 articles from same box give one defective. So $P(E\,|\,A)$ etc are binomial. Answer is $\frac{162}{1205}$.

12. Do as $P(1,0,0) + P(0,1,0) + P(0,0,1)$ etc. Answers are $\frac{426}{1000}$, $\frac{549}{2500}$.

13. Part (b) is similar to Ex.12. Answers are (a) $\frac{243}{1000}$, (b) $\frac{283}{1000}$. Last part: first work out the probability that a random chair is defective as in §22. Answer = $\frac{4 \cdot 13^2 \cdot 87^3}{10^{10}}$.

EXAMPLES 7 (PAGES 50-51)

1. (i) $\frac{4}{27}$, (ii) $\frac{1}{9}$, (iii) $\frac{7}{48}$.

2. (i) $\frac{1}{80}$, (ii) $\frac{39}{40}$, (iii) $\frac{6}{17}$. Last part is binomial: answer = $\frac{54 \cdot 17^2}{20^4} = 0.0975$.

3. (i) $\frac{1}{64}$, (ii) $\frac{225}{256}$, (iii) $\frac{144}{225}$. For the last part write out the possible sequences and add up their probabilities. Answer = $\frac{31^2 \cdot 225^2 \cdot 481}{256^5}$ = 0.0213.

4. (i) $F(x) = 0$ $(x < 4)$ and $F(x) = 1 - \frac{4}{x}$ $(x \geq 4)$, (ii) $\frac{2}{3}$, (iii) $\frac{2}{5}$.

EXAMPLES 8 (PAGES 64-67)

1. (a) 13, (b) 17.

3. (a) $\frac{2}{3}, \frac{1}{18}$, (b) $\frac{1}{2}, \frac{1}{20}$, (c) $\frac{3}{5}, \frac{1}{25}$, (d) $\frac{b+a}{2}$, $\frac{(b-a)^2}{12}$, (e) Gamma function from §100 helps. Answers: $\frac{1}{a}, \frac{1}{a^2}$. (f) $\frac{2}{a}$, $\frac{2}{a^2}$ (Gamma function helps.)

4. 2, $\frac{2\sqrt{3}}{3}$. Last part: No. See the last two sentences in §56.

5. £41.92.

6. £(2B-3A)/3.   7. $5 - 3e^{-1/3} - e^{-2/3} - e^{-1}$.

8. £$(2 + e^{-1} - 2.5e^{-2})$.   9. £$\left(B - A - \frac{CD}{1000}\right)$.

10. Write out a money taken in function $M(t)$ rather as in §55, but notice $M(t)$ involves t, e.g. $M(t) = tB$ for $2 \leq t \leq 3$, etc. Then find $E(M(t))$ using part (b) of the result in §52. Answer is £$\left(\frac{8B + 4C}{3} - A - 50\right)$.

11. £$\left(\frac{1}{2}B(e^{-1} + e^{-\frac{1}{2}}) - A\right)$. Last bit is conditional probability with answer $1 - e^{-\frac{1}{2}}$.

12. 1, $\frac{1}{2}$ for X; $\frac{3\sqrt{\pi}}{4}$, $\frac{32-9\pi}{16}$ for Y. The result on the Gamma function from §100 is a help here.

13. $E(X) = 200$, s.d. = 10. (Binomial: use §63.)

14. Find $E(X^2)$ with part (a) of the result in §52. Answers are (a) 17.31, (b) 12.20.

15. £20.55.

16. Write out the cases and find the probabilities of each total score. $E(X) = 7$, s.d. = 2.415.

18. Binomial: use §63. $E(X) = 75$, s.d. = 7.5.

EXAMPLES 9 (PAGES 76-78)

1. (a) $e^{-3} = 0.050$, (b) $1 - \frac{17}{2}e^{-3} = 0.577$, (c) $\frac{12}{e^3 - 1} = 0.629$.

2. (a) $\frac{4e^{-2}}{3} = 0.180$, (b) $1 - 5e^{-2} = 0.323$. (c) You need §74. Answer = $50e^{-10} = 0.00227$. (d) Binomial. Answer = $5e^{-8}(1 - e^{-2}) = 0.00145$.

3. (a) Binomial: 0.349, 0.387, 0.194, 0.057.
   (b) Poisson: 0.368, 0.368, 0.184, 0.061.

4. Like §72. (a) $\frac{9}{2}e^{-3} = 0.224$, (b) $1 - 4e^{-3} = 0.801$. Last part = $1 - 3.4e^{-2.4} = 0.692$.

5. (a) $18e^{-6} = 0.045$, (b) $1 - 61e^{-6} = 0.849$, (c) $\frac{918}{1223} = 0.751$.

6. $1 - e^{-1} = 0.632$. Second part - use either binomial with $n = 40$, $p = 0.001$ or Poisson approximation with $\lambda = 0.04$. Answer (either way) is 0.9608. Last part: $n = 100$, $p = 0.0392$. Answer is 0.750.

7. You need §74 here. (a) 0.0150, (b) 0.00623.

8. You need to write out the expected pay out in series form with part (a) of the result in §52. The proof in §68 will help to sum the series. The answers are (a) £3, (b) £7.50.

EXAMPLES 10 (PAGES 87-89)

WARNING: The $N(0,1)$ table is given to 4 decimal places and the answers here are also given to 4 decimal places on the basis of the values in the table. These answers can be regarded as correct to 4 decimal places if they involve reading only one value from the table. If however two values from the table have been subtracted (as in $\Phi(b) - \Phi(a)$) or added or multiplied or if a value from the table has been raised to a power (as in the binomial) then inaccuracy can creep in and the fourth decimal place could be wrong. To illustrate, suppose two exact Φ values are 0.60046 and 0.60013. The table would show these as 0.6005 and 0.6001, and if we subtracted these to give an answer we would obtain 0.0004. The true value, however, is 0.60046 - 0.60013, i.e. 0.00033 = 0.0003 (to 4 decimal places).

1. (a) 0.3413, (b) 0.2117, (c) 0.5160, (d) 0.0668, (e) 0.0228, (f) 0.4209.

2. (a) 0.4013, (b) 0.7888, (c) 0.7369. Last part is binomial with answer 0.2892.

3. (a) 0.2902, (b) 0.5987, (c) 0.1764.

4. (a) 0.2268, (b) 0.2902. The third part is binomial with answer 0.1807. Fourth part = 0.0125.

5. English is better, i.e. higher in $N(0,1)$.

6. (a) 0.2119, (b) 0.6826, (c) 0.5432. For the last part it is probably easier to find which bulb is more likely to fail before 230 hours and then choose the other one. The first type is better.

7. (a) 0.1151, (b) 0.5138, (c) 0.3656. Shipyard B is more likely to complete it in the time.

8. 0.4444, 0.1556.

9. (a) A, (b) B, (c) B. For the last part work out $1 - P$(neither house completed), i.e. work out $1 - P(A > 122$ and $B > 122)$. Answer = 0.3370.

10. Copy the example in §84. You will need to interpolate in the table. Answer = 0.666.

11. Copy the example in §84. You will need to interpolate in the table. Answer = 0.314.

12. Use the result of §84. Each is $N(50,400)$. The answers are (a) 0.059, (b) 0.132.

EXAMPLES 11 (PAGES 96-97)

1. (a) $f(x,y) = \frac{1}{16}$ ($0 \le x \le 4$, $0 \le y \le 4$), $f(x,y) = 0$ (otherwise). (b) $\frac{3}{4}$, (c) $\frac{31}{32}$, (d) $\frac{23}{31}$.

2. (a) $\frac{1}{72}$, (b) $\frac{5}{36}$, (c) $\frac{1}{216}$.

3. (a) $\frac{1}{8}$, (b) $\frac{1}{4}$, (c) The event is given by $0 \le x \le 1$ and $0 \le y \le 1$. Answer = $\frac{1}{32}$. (d) $\frac{21}{32}$, (e) Use (c). Answer = $\frac{31}{32}$. (f) Use (d). Answer = $\frac{11}{32}$.

4. The result in §100 helps in the first part. Answer = $\frac{9}{49}$. Second part needs integration by parts. Answer is $1 - 4e^{-3} - 9e^{-4}$.

5. e^{-1}. (Note that the range for y is $(0,1)$.)

6. (a) $1 - 5e^{-4}$, (b) $\frac{7}{27}$. Last part is binomial with answer $1 - (1-q)^{12}$.

7. For $E(X)$, $V(X)$ see §61. Last part = $\frac{1}{5}$.

8. The result in §100 on the Gamma function will help here. Answer = $\frac{8}{125}$.

EXAMPLES 12 (PAGES 107-108)

1. Similar to §99. Here $E(X) = 800$ and the standard deviation is 20. Remember the correction for continuity. Answers are (a) 0.9570, (b) between 0.9999 and 1, (c) between 0.9999 and 1 but closer to 1 than (b). Part (b) is of interest to the person who tosses 703 heads in the first sentence of §94. Chebyshev's lower bounds are pitifully short in this problem.

2. Similar to §99. $E(X) = 30$ and the standard deviation is 5. Remember the correction for continuity. Answers are (a) $1 - \Phi(2.9) = 0.0019$, (b) 0.7286. Chebyshev as we stated it can only contribute to (b). However the range from 24.5 to 35.5 corresponds to $k = 1.1$ in Chebyshev and this gives little leverage: remember we remarked in the second sentence on page 99 that $k = 1$ is hopeless. Actually putting $k = 1.1$ gives $P(25 \leq X \leq 35) \geq 0.174$.
Writing a computer program to find $P(25 \leq X \leq 35)$ exactly by finding the sum $P(X=25) + P(X=26) + \ldots + P(X=35)$, where each term is calculated with the binomial formula of §37, gives the exact value of $P(25 \leq X \leq 35)$ as 0.7292. In writing your program be careful about the order in which you multiply factors together or you will have problems with overflow.

5. Similar to §99. $E(X) = 12$ and the standard deviation is 3. Remember the correction for continuity. Answers are (a) 0.132, (b) 0.866, (c) 0.067. [In (b) the range is taken from 7.5 to 16.5, while in (c) it is taken from 16.5 upwards.] You could write a program as in Ex. 2 above to check the accuracy of the approximations.

6. Take the normal approximation to the binomial as in §99 with $n = 100$ and $p = 0.9$. We want $P(X > 95.5)$. Answer = 0.033.

7. For example let the random variable X take the values $-3, 0, 3$ with probabilities $1/18, 8/9, 1/18$.

TABLE OF CUMULATIVE DISTRIBUTION FUNCTION Φ
FOR THE STANDARD NORMAL DISTRIBUTION N(0,1)

(See §78 for the definiton of Φ)

a	Φ(a)	a	Φ(a)	a	Φ(a)	a	Φ(a)
0.00	0.5000	0.58	0.7190	1.16	0.8770	1.74	0.9591
0.01	0.5040	0.59	0.7224	1.17	0.8790	1.75	0.9599
0.02	0.5080	0.60	0.7257	1.18	0.8810	1.76	0.9608
0.03	0.5120	0.61	0.7291	1.19	0.8830	1.77	0.9616
0.04	0.5160	0.62	0.7324	1.20	0.8849	1.78	0.9625
0.05	0.5199	0.63	0.7357	1.21	0.8869	1.79	0.9633
0.06	0.5239	0.64	0.7389	1.22	0.8888	1.80	0.9641
0.07	0.5279	0.65	0.7422	1.23	0.8907	1.81	0.9649
0.08	0.5319	0.66	0.7454	1.24	0.8925	1.82	0.9656
0.09	0.5359	0.67	0.7486	1.25	0.8944	1.83	0.9664
0.10	0.5398	0.68	0.7517	1.26	0.8962	1.84	0.9671
0.11	0.5438	0.69	0.7549	1.27	0.8980	1.85	0.9678
0.12	0.5478	0.70	0.7580	1.28	0.8997	1.86	0.9686
0.13	0.5517	0.71	0.7611	1.29	0.9015	1.87	0.9693
0.14	0.5557	0.72	0.7642	1.30	0.9032	1.88	0.9699
0.15	0.5596	0.73	0.7673	1.31	0.9049	1.89	0.9706
0.16	0.5636	0.74	0.7704	1.32	0.9066	1.90	0.9713
0.17	0.5675	0.75	0.7734	1.33	0.9082	1.91	0.9719
0.18	0.5714	0.76	0.7764	1.34	0.9099	1.92	0.9726
0.19	0.5753	0.77	0.7794	1.35	0.9115	1.93	0.9732
0.20	0.5793	0.78	0.7823	1.36	0.9131	1.94	0.9738
0.21	0.5832	0.79	0.7852	1.37	0.9147	1.95	0.9744
0.22	0.5871	0.80	0.7881	1.38	0.9162	1.96	0.9750
0.23	0.5910	0.81	0.7910	1.39	0.9177	1.97	0.9756
0.24	0.5948	0.82	0.7939	1.40	0.9192	1.98	0.9761
0.25	0.5987	0.83	0.7967	1.41	0.9207	1.99	0.9767
0.26	0.6026	0.84	0.7995	1.42	0.9222	2.00	0.9772
0.27	0.6064	0.85	0.8023	1.43	0.9236	2.05	0.9798
0.28	0.6103	0.86	0.8051	1.44	0.9251	2.10	0.9821
0.29	0.6141	0.87	0.8078	1.45	0.9265	2.15	0.9842
0.30	0.6179	0.88	0.8106	1.46	0.9279	2.20	0.9861
0.31	0.6217	0.89	0.8133	1.47	0.9292	2.25	0.9878
0.32	0.6255	0.90	0.8159	1.48	0.9306	2.30	0.9893
0.33	0.6293	0.91	0.8186	1.49	0.9319	2.35	0.9906
0.34	0.6331	0.92	0.8212	1.50	0.9332	2.40	0.9918
0.35	0.6368	0.93	0.8238	1.51	0.9345	2.45	0.9929
0.36	0.6406	0.94	0.8264	1.52	0.9357	2.50	0.9938
0.37	0.6443	0.95	0.8289	1.53	0.9370	2.55	0.9946
0.38	0.6480	0.96	0.8315	1.54	0.9382	2.60	0.9953
0.39	0.6517	0.97	0.8340	1.55	0.9394	2.65	0.9960
0.40	0.6554	0.98	0.8365	1.56	0.9406	2.70	0.9965
0.41	0.6591	0.99	0.8389	1.57	0.9418	2.75	0.9970
0.42	0.6628	1.00	0.8413	1.58	0.9429	2.80	0.9974
0.43	0.6664	1.01	0.8438	1.59	0.9441	2.85	0.9978
0.44	0.6700	1.02	0.8461	1.60	0.9452	2.90	0.9981
0.45	0.6736	1.03	0.8485	1.61	0.9463	2.95	0.9984
0.46	0.6772	1.04	0.8508	1.62	0.9474	3.00	0.9987
0.47	0.6808	1.05	0.8531	1.63	0.9484	3.10	0.9990
0.48	0.6844	1.06	0.8554	1.64	0.9495	3.20	0.9993
0.49	0.6879	1.07	0.8577	1.65	0.9505	3.30	0.9995
0.50	0.6915	1.08	0.8599	1.66	0.9515	3.40	0.9997
0.51	0.6950	1.09	0.8621	1.67	0.9525	3.50	0.9998
0.52	0.6985	1.10	0.8643	1.68	0.9535	3.60	0.9998
0.53	0.7019	1.11	0.8665	1.69	0.9545	3.70	0.9999
0.54	0.7054	1.12	0.8686	1.70	0.9554	3.80	0.9999
0.55	0.7088	1.13	0.8708	1.71	0.9564	3.90	1.0000
0.56	0.7123	1.14	0.8729	1.72	0.9573	4.00	1.0000
0.57	0.7157	1.15	0.8749	1.73	0.9582		

INDEX

(The numbers refer to sections <u>not</u> to pages.)

Ancestry, 23.
Average of observations, 65.

Bayes' theorem, 23.
Binomial coefficients, 11.
Binomial distribution, 36, 63.

Central Limit Theorem, 95.
Chebyshev's Inequality, 92.
Circuits, 31.
Combinations, 11.
Complement, 1.
Conditional probability, 16.
Continuous, 34, 41.
Correction for continuity, 97, 98, 99.
Countably infinite, 34.
Cumulative distribution function, 43.

De Morgan's Laws, 1.
Discrete, 34.
Disjoint, 5.
Distributive Laws, 1.
Dwindling population, 40.

Empty set, 1.
Equally likely outcomes, 13.
Event, 4.
Evidence, 23.
Expected value, 47 - 56.
Exponential distribution, 45.

Factorial, 9.

Gamma function, 100.

Hypergeometric, 14, 40.

Independence, 25 - 30, 88.
Intersection, 1.

Joint probability density function, 87.

Mean, 47 - 56.
Multiplication principle, 8.
Mutually exclusive, 5.

Normal approximation to binomial, 98 - 99.

Normal distribution, 76 - 84.

Order is important, 10.
Order is unimportant, 11 - 12.
Outcome set, 3.

Partition, 22.
Permutations, 9.
Poisson, 66 - 75.
Poisson approximation to the binomial, 70 - 73.
Probability density function, 41.
Probability of an event, 6.
Production line, 40.

Random experiment, 2.
Random variable, 33 - 35.
Record set, 3.
Replacement, 15.

Sample space, 3.
Sand, 6.
Set theory, 1.
Standard deviation, 58.
Standard normal distribution, 77.
Subset, 1.
Supply and demand problems, 91.

Two dimensional random variables, 85.

Uniform distribution, 46.
Union, 1.

Variance, 47, 57 - 62.
Venn diagram, 1.

$\emptyset, \cup, \cap$: See §1.
$x \in A$ means x is a member of set A.
$N(\mu, \sigma^2)$: See §76.
Φ function: See §78.
The number e has the value 2.71828... .

ALSO AVAILABLE IN THIS SERIES ARE THE FOLLOWING:

ADVANCED CALCULUS FOR ENGINEERING AND SCIENCE STUDENTS

Ian S. Murphy.   Third Edition, 1989.   248 pages.

Arklay Publishers, 64 Murray Place, Stirling.

Contents: Double and Triple Integration, Beta and Gamma Functions, Differential Equations, Laplace Transforms, Partial Differentiation, Errors and Exact Differentials, Vector Calculus, Line and Surface Integrals, Fourier Series, Maxima and Minima of Functions of Several Variables, Eigenvalues and Eigenvectors of Matrices, Systems of Linear Differential Equations, Lagrange Multipliers, Hints and answers to the examples, Index.

BASIC MATHEMATICAL ANALYSIS: THE FACTS

Ian S. Murphy.   Third Edition, 1991.   246 pages.

Arklay Publishers, 64 Murray Place, Stirling.

Contents: Basic ideas, Limits of sequences, Limits of real functions, Continuity, Metric spaces and compactness, Derivatives and applications, Series, Taylor and Maclaurin expansions, More on series, Power series, Exponential, logarithmic, trigonometric and hyperbolic functions, Integration, More on the convergence of sequences and series, Miscellaneous examples, Hints and answers to examples, Index.